ROUTLEDGE LIBRARY EDITIONS:
THE NINETEENTH-CENTURY NOVEL

Volume 33

GEORGE GISSING

GEORGE GISSING
Voices of the Unclassed

Edited by
MARTIN RYLE AND JENNY BOURNE TAYLOR

Routledge
Taylor & Francis Group
LONDON AND NEW YORK

First published in 2005 by Ashgate Publishing Ltd.

This edition first published in 2016
by Routledge
2 Park Square, Milton Park, Abingdon, Oxon OX14 4RN

and by Routledge
711 Third Avenue, New York, NY 10017

Routledge is an imprint of the Taylor & Francis Group, an informa business

British Library Cataloguing in Publication Data
A catalogue record for this book is available from the British Library

ISBN: 978-1-138-67777-7 (Set)
ISBN: 978-1-315-55928-5 (Set) (ebk)
ISBN: 978-1-138-67552-0 (Volume 33) (hbk)
ISBN: 978-1-138-67553-7 (Volume 33) (pbk)
ISBN: 978-1-315-56065-6 (Volume 33) (ebk)

Publisher's Note
The publisher has gone to great lengths to ensure the quality of this reprint but points out that some imperfections in the original copies may be apparent.

George Gissing

Voices of the Unclassed

Edited by
MARTIN RYLE
University of Sussex, UK

and

JENNY BOURNE TAYLOR
University of Sussex, UK

ASHGATE

Published by
Ashgate Publishing Limited
Gower House
Croft Road
Aldershot
Hants GU11 3HR
England

Ashgate Publishing Company
Suite 420
101 Cherry Street
Burlington, VT 05401-4405
USA

Ashgate website: http://www.ashgate.com

British Library Cataloguing in Publication Data
George Gissing : voices of the unclassed.—(The nineteenth cenury series)
1.Gissing, George, 1875-1903—Criticism and interpretation I.Ryle, Martin H.
II.Taylor, Jenny Bourne, 1949-
823.8

Library of Congress Cataloging-in-Publication Data
George Gissing : voices of the unclassed / edited by Martin Ryle and Jenny Bourne Taylor.
p. cm.— (The nineteenth century series)
Includes bibliographical references.
ISBN 0-7546-3675-5 (alk. paper)
1. Gissing, George, 1857-1903—Criticism and interpretation. 2. Gissing, George, 1857-1903—Political and social views. 3. Literature and society—England—History—19th century. 4. Social problems in literature.
I. Ryle, Martin H. II. Taylor, Jenny Bourne, 1949- III. Series: Nineteenth century (Aldershot, England)

PR4717.G45 2005
823'.8—dc22

2004026339

ISBN 0 7546 3675 5

Printed and bound in Great Britain by MPG Books Ltd, Bodmin, Cornwall

Contents

Notes on Contributors

David Glover is Senior Lecturer in English at the University of Southampton, where he teaches late Victorian and Edwardian literature, Irish literature, and cultural theory. His recent books include *Vampires, Mummies, and Liberals: Bram Stoker and the Politics of Popular Fiction* (1996) and *Genders* (2000; with Cora Kaplan). He is currently working on a cultural history of the 1905 Aliens Act.

Dr Simon J. James is Lecturer in Victorian Literature at the University of Durham, where he teaches mainly nineteenth- and twentieth-century literature and critical theory. He is the author of *Unsettled Accounts: Money and Narrative in the Novels of George Gissing* (Anthem, 2003), and of other articles on Gissing. He has edited Gissing's *Charles Dickens: A Critical Study* (Grayswood, 2004) and four H.G. Wells novels for Penguin Classics.

Dr Emma Liggins lectures in Victorian literature at Edge Hill College of Higher Education. She has published articles on the New Woman and sensation fiction in *Women's Writing*, *Literature and History* and *Journal of Victorian Culture*. She has also edited (with Daniel Duffy) *Feminist Readings of Victorian Popular Texts: divergent femininities* (Ashgate, 2001). She has a chapter on Gissing and social investigation in *Women's Work Cultures, 1850-1950* ed. Louise Jackson & Krista Cowman (Ashgate, 2005). She is the author of *George Gissing, the Working Woman and Urban Culture* (Ashgate, 2005).

Christina Lupton teaches at The University of British Columbia, Vancouver. Her work focuses on eighteenth- and nineteenth-century literature, aesthetic theory, and the question of self-consciousness in fiction. She has published articles in *English Literary History*, *Modern Language Notes*, and *New German Critique*.

Diana Maltz is an Assistant Professor at Southern Oregon University where she teaches classes in nineteenth-century literature and culture. Her research focuses on the intersections between late-Victorian poverty, aesthetics, and social reform. She has published essays on George Gissing, Oscar Wilde, housing reformer Octavia Hill and aesthetician Vernon Lee. Her book *British Aestheticism and the Urban Working Classes, 1875-1900: Beauty for the People* is forthcoming from Palgrave in October, 2005.

Scott McCracken is Professor in the School of English at Keele University. He is the author of *Pulp: Reading Popular Fiction* (Manchester UP, 1998) and the editor of the journal *new formations.* He is currently completing a book, *Urban Masculinities and Modernist Fiction,* for Manchester University Press.

Patrick Parrinder is a Professor of English at the University of Reading. His books include studies of James Joyce, H.G. Wells, science fiction, and literary criticism. He is currently writing a history of the novel and English national identity, provisionally titled *Fictions of the Nation,* for Oxford University Press. He was an associate editor of the Oxford Dictionary of National Biography with responsibility for literary authors 1890-1920, and he is general editor of the new editions of H. G. Wells recently published in Penguin Classics.

Deborah Parsons is a Senior Lecturer in Nineteenth and Twentieth-Century English Literature at the University of Birmingham. She is the author of *Streetwalking the Metropolis* (OUP, 2000), *Djuna Barnes* (Northcote House, 2003), *A Cultural History of Madrid: Modernism and the Urban Spectacle* (Berg, 2003), and various essays on urban culture and literary modernism.

Tilman Reitz teaches philosophy at the Friedrich-Schiller-University of Jena, Germany. He is working in the fields of social theory, aesthetics, and political epistemology. Recent publications include his dissertation, *Bürgerlichkeit als Haltung. Zur Politik des privaten Weltverhältnisses* (Fink: Munich 2003), and various articles, e.g. on Spinoza, Adorno, Foucault, and Derrida, most of them in the review *Das Argument.*

Martin Ryle is Senior Lecturer in Continuing Education at the University of Sussex, where he teaches literature and cultural studies. His books include *Journeys in Ireland: Literary travellers, rural landscapes, cultural relations* (Ashgate, 1997) and, with Kate Soper, *To Relish the Sublime? Culture and Self-Realization in Postmodern Times* (Verso, 2002). His main current interest is in changing modes of historical representation and consciousness in the novel.

Jenny Bourne Taylor is a Reader in English at the University of Sussex. She has published widely on nineteenth-century literature and culture, in particular on nineteenth-century psychology and its relation to literature. Publications include *In the Secret Theatre of Home: Wilkie Collins, Sensation Narrative and Nineteenth-Century Psychology* (Routledge 1988) and (edited with Sally Shuttleworth) *Embodied Selves: an Anthology of Psychological Texts* (OUP 1998). Her current interests are in concepts of illegitimacy in law and literature, and notions of agency in nineteenth-century psychology and literature.

Introduction: Gissing's Critical Contexts

Martin Ryle and Jenny Bourne Taylor

George Gissing, who was born in the heyday of the Victorian era and died at the dawn of the twentieth century, is now seen as an integral part of the cultural landscape of the *fin de siècle*, but his position within this landscape remains a peculiar one. Unlike Hardy, say, or Conrad, whose place on the cusp of modernism also rested on their marginality, his edgy position as the voice of the 'unclassed' is as shifting and unstable as the social and cultural identities and tensions that he dramatizes. Gissing, perhaps more than any late nineteenth-century writer, has attracted a particular kind of biographical criticism – and this is not surprising given how the experience of 'exile', and the liminal position of the intellectual with which he grappled during his life, lie at the narrative centre of so many of his novels. At the same time his profile has taken on new shapes for successive generations of critics, and the way in which he speaks to the readers who have rediscovered and reinvented him maps the history of cultural criticism and theory over the last thirty years, stretches its boundaries and exposes some of its crucial contours. Gissing himself stressed the need to historicize the conditions which shape a writer's work, position and 'public', both in *New Grub Street* and in his study of Dickens, and while critics have highlighted the contexts – cultural, political and discursive – that shape the parameters of his writing and the paradigms for reading it, he does not passively fit into any of them. For Gissing intervenes, speaks backs to his readers as he dramatizes the successive cultural and social 'crises' of the 1880s and 90s: crises of class relations, culture, and social transformation; of new patterns and technologies of literary production as an aspect of consumer capitalism; of gender divisions and relationships; of late imperialism and global capitalism. Here we will not attempt to follow all aspects of that conversation, but highlight some key moments, starting with the 1960s and 1970s, before we introduce the essays that make up the present volume.

Gissing in the 1960s and 1970s

That conversation probably would not have taken place at all if Pierre Coustillas and John Spiers had not made the reprinting of Gissing's novels one of the linchpins of the Harvester Press – a new independent publisher, founded in 1970, that went on to play a significant part in the transformation of literary studies through the two following decades. The Harvester project made Gissing's fiction physically available to a wider readership and played a crucial role in shaping his position as both a complex realist and a self-reflexive cultural critic who actively engaged in the cultural debates dramatized in his fiction – though Gissing, characteristically, never quite occupied the space marked out for him by the project (Coustillas and Spiers, 1971). This was not a new venture – Gissing had already figured in the development of cultural analysis and Victorian studies as an interdisciplinary and international enterprise from the late 1950s. Jacob Korg's biography had not only revised many of the inaccuracies of Morley Roberts's thinly disguised *The Private Life of Henry Maitland* of 1912, but set a standard for biographically inflected criticism in the 1960s that would be exploited by Gillian Tindall in the 1970s and extended by John Halperin in the 1980s (Korg, 1963; Tindall, 1975; Halperin, 1982).

Indeed, Raymond Williams allotted Gissing a key place in the 'Interregnum' – that uneasy hinterland in the aftermath of high Victorianism – in *Culture and Society* in 1958, precisely because his early work embodied Williams' own specific preoccupations with the active role of 'culture' in shaping class relations. For Williams, Gissing stands somewhere between Gaskell and Orwell, but he also marks the shift, still implicit in *Culture and Society*, between 'culture' itself as an idea and as a material practice. His novels of social crisis, such as *Demos*, *Thyrza* and *The Nether World*, Williams argues, 'provide evidence of a significant and continuing process' that continues the debates of the condition of England novels of the 1840s by marking the limits of their meliorist concept of culture and community; while *New Grub Street* articulates and analyses a precise moment in the commodification of culture that is equally central to its transformations through the nineteenth and early twentieth centuries (Williams, 1958, p.175).

Raymond Williams read Gissing primarily as a 'social problem' novelist of the 1880s; but he also highlighted Gissing's critical analysis of what he termed 'negative identification', his engagement with the necessary relationship between the personal and the political (as the 1960s would have it), even if that relationship might be a hostile one. This projective identification is voiced by Osmond Waymark in *The Unclassed* – ' "That zeal on behalf of the suffering masses was nothing more nor less than disguised zeal on behalf of my own starved passions.... I identified myself with the poor and ignorant; I did not make their cause my own but my own cause theirs" ' – and it is one that leads to the desire for disengagement and, in turn, the realization of the impossibility of autonomy. It was this more symptomatic and dynamic sense of identity, ideology and intellectual intervention that was central to Gissing studies of the 1960s and 70s, and the contrasting directions that Williams' analysis could take can be traced in John Goode and Adrian Poole's work.

In his discussions of Gissing in the 1960s, Goode like Williams focused on the 1880s, in particular on *The Nether World* and on *Demos: A Story of English Socialism* (Howard, Lucas and Goode, 1966; Goode, 1968). His emphasis, however, is on the constraints and possibilities inherent within different forms of realism, and on the precise political circumstances in which Gissing's interventions take place – above all the debates and divisions within the socialist movement. The figure of William Morris is central here, as a political and cultural presence that acts as both counterpoint and framework to Gissing's own shifting ideological position. Both writers rejected the reformist progressivism of Positivism in the early 1880s, but although this rejection lead them in contrary directions (Gissing to Schopenhauerian withdrawal, Morris to more active political engagement in the Socialist League) Goode places them at opposing ends of a shared political dynamic. In *The Nether World*, Gissing's ambiguous class position means that his pessimism feeds a critical realism (in Lukács' sense) that is able to represent the complete breakdown of the social structure: the world he portrays is not like hell, it *is* hell. It is a world that Morris used a utopian 'culture' to escape from and to imagine a possible alternative – a possibility open to him precisely because of his material and cultural privilege – but it is Gissing's nightmare that generates a compelling historical energy, rather than Morris's dream of passive perfection.

Goode's discussion of *Demos*, the fable of a working-class socialist who unexpectedly inherits a fortune and his doomed attempt to set up a socialist community run on broadly Owenite lines, takes the analysis of the interpretive possibilities within Gissing's explicit conservatism further. Here Goode inverts Morris's implied interventionist role, for while Morris provided an analytical framework for *The Nether World*, in *Demos* he became a figure within the text itself. *Demos* was conceived as a 'savage satire on working class aims and capacities', and was incorporated into anti-socialist discourse during the 1880s; but Goode shifts the emphasis from reading the novel as an instructive breakdown of negative identification to stressing the novel's mimetic possibilities within its precise political context. Arguing that '*Demos* is confused partly because the situation, and the general consciousness of that situation, are confused. Do we expect a novel to transcend the limitations of its epoch?' (1968, p.202), Goode teases out the different processes which converged on the novel's publication, and excavates a series of realisms which may be at odds with each other, and which refuse to fit into any simple division between text and context. There is Gissing the 'writer as producer' in Benjamin's terms, who was at a particular crisis of his career when he published *Demos*, aiming it (successfully) at the 'Mudie' market. The sentimental conservatism of this readership, though, is undercut by Gissing the implied author, who draws on the contradictory discourses of Spencer's individualist evolutionism and Schopenhauer's image of the alienated self against a destructive world in reacting against his earlier adherence to a radical Positivism. This self-undermining process is set against the external context of *Demos*, and here Goode stresses the novel's mimetic accuracies and inaccuracies, in which the documentary realism of Hoxton streets contrasts with a far more equivocal portrayal of the socialist movement itself. Again these tensions come to head in the figure of Westlake, the novel's fictionalized Morris (Westlake is a parody of Morris's radical aestheticism); but, in contrast to *The Nether World*, the novel itself dramatizes its own theory of artistic production, in which a socialist consciousness can only exist as a dreamlike possibility, rather than though a destabilized realism.

John Goode's discussion of how to read a text within its immediate political 'context' proved controversial, but his dense analysis demonstrated how a new kind of theoretical awareness could be used to tease out its ideological connotations without sacrificing narrative complexity. His full-length study, *George Gissing, Fiction and Ideology* (1979), anticipated the discussion of the discursive mapping of Gissing's fiction that was taken up by Jameson in the *The Political Unconscious;* but it also developed and responded to Adrian Poole's important study of 1975, *Gissing in Context*, and the two books together demonstrate the sophistication of Gissing studies by the end of the 1970s.

If Goode's focus was on 'ideology', with all the complexity that the term would acquire in the 1980s, Poole's focus was on 'culture'. He develops Williams' analysis of the intricate historical process surrounding late nineteenth-century cultural production by setting Gissing's work within the intersecting environments that it transects: shifts in the nineteenth-century novel and its position; representations of the city; the 'massification' of culture and the place of the writer; constructions of gender and debates on marriage. Within these settings *Gissing in Context* takes 'negative identification' and its implications for identity-formation itself far beyond Williams' framework, implicitly anticipating some of the issues raised by deconstruction and psychoanalysis, even if not exactly in their terms: Poole is concerned above all with the historicized self and its contradictions. He follows through Gissing's own analysis of the psychology of 'exile', arguing that his narratives are driven, phenomenologically, by a dynamic of ambivalent desire and its negation which generates and is structured through a series of oppositions – between self and other, artist and society, autonomy and connection, public and private worlds – oppositions which are continually breaking down and regenerating but never resolved. This is a dynamic of the *fin de siècle*, and is part of the wider sense of 'cultural crisis' of the late nineteenth century, its preoccupation with shoring up boundaries rather than crossing them; but Gissing was less 'well padded and bolstered' than his contemporaries and this means that he not only represents the erosion of earlier Victorian ideals and aspirations more starkly, but paradoxically marks their residual traces.

Like Goode, Poole is concerned with Gissing's extraordinary ability to articulate the very ideological contradictions that he remains defined by and trapped within – with the capacity of his narratives to transcend their own conditions of production and legibility. However, his stress is on this dynamic within the self, and with making the texts coherent within the terms of that dynamic, whereas Goode insists that Gissing ultimately refuses epistemological and ideological coherence. The contrast can be seen in their discussions of Godwin Peak's situation as an intellectual in *Born in Exile*. For Poole, the novel marks the culmination of the tension between the desire for connection and autonomy: it is the inner desire for assimilation as much as outer 'respectability' that lies behind Peak's drive to succeed. It is in his inevitable failure to reject his rational secularism, to lead a double life 'in which public performance and inner feelings are at total variance', that the novel's radical potential lies: the 'personal' defines the political (Poole, 1975, p.171). For Goode, in contrast, *Born in Exile* exposes the ideological contradictions of the intellectual in Gramsci's terms – as a member of a group with a specific social function – and this shapes Peak's conflicted identity.

Gissing and criticism since 1980

Since Gissing's oeuvre began to be republished in the 1960s, his work has been discussed in several critical monographs devoted more or less exclusively to him –

including those by John Goode and Adrian Poole discussed above, and important later studies by David Grylls, John Sloan and Robert L. Selig. His fiction has also been analysed in more general literary histories and interdisciplinary cultural studies: for instance, he is an important figure, at once representative and original, in John Gross's The Rise and Fall of the Man of Letters, and is discussed in William Greenslade's Degeneration, culture and the novel 1880–1940 and Daniel Born's The Birth of Liberal Guilt in the English Novel. Other discussions of his work in the last two decades include those by Fredric Jameson, Rachel Bowlby and John Carey which we shall briefly consider; and also, among others, P.J. Keating's The Haunted Study (1989), Patricia Alden's Social Mobility in the English Bildungsroman: Gissing, Hardy, Bennett and Lawrence (1986) and Sally Ledger's The new woman: fiction and feminism at the fin de siècle (1997). There is also an extensive critical literature in periodicals: the Gissing Newsletter and Gissing Journal, and journals dealing more generally with late nineteenth-century writing and cultural history. The International George Gissing Conference held in Amsterdam in September 1999, from which many contributions were subsequently printed in A Garland for Gissing (Postmus, 2001), indicated the range of perspectives from which specialist work on Gissing is now undertaken. In 2003, a major conference was held in London on the theme of Gissing and the City. We should mention, finally – though it is not our present concern – the biographical and bibliographical research which has enabled publication of scholarly editions of Gissing's correspondence and other papers and is reflected in a number of studies of his life, culminating in a complete collection of his letters (Mattheisen, Young and Coustillas, 1990–7).

Fredric Jameson's account of Gissing in The Political Unconscious: Narrative as a Socially Symbolic Act (Jameson, 1981, pp.185-205) takes up the questions of ideology, representation and narrative form that had concerned Goode. Jameson focuses initially on Gissing's reliance, in The Nether World, on 'ideologemes' drawn from the Dickensian repertoire. (Ideologemes have earlier been defined as narrative paradigms that embody ideas, or 'pseudoideas', concerning class relations.) He argues that neither this Dickensian blend of melodrama and sentiment nor the more Zola-like 'systematic mapping' which supplements it offers a workable solution to the problems of representation that the novel confronts. The Nether World seeks to pass itself off as a 'map of the social totality', but depends on a 'privileged but placeless observer', ineluctably caught up in the class relations which this imaginary placelessness seeks to disavow.

Jameson goes on to chart the mutation of the 'placeless observer' into the 'alienated intellectual', and to consider how in his 'greatest novels' Gissing embodies intellectual alienation in the specific figure of the writer. More generally, he sees the later fiction as deploying a characterological 'system [which] can be understood only in terms of desire', while representing a world in which all desires have been commodified: consequently, 'any achieved desire' is 'inauthentic, while an authenticity at best pathetic clings to images of failure'. Gissing reveals 'historical and social reality' by 'generating an omnipresent class consciousness in which it is intolerable for the bourgeois reader to dwell for any length of time' (ibid., pp.204f.). Hence, Jameson concludes, the justification for hailing Gissing as the writer of 'authentic ressentiment'. This complex argument, although based on a highly selective sampling of the novels, illuminates what many readers will experience in reading Gissing: the sense that he remains deeply mired in the very conflicts that he seeks to chart and master, and that precisely this tension is what distinguishes his best work. The chapter confirms, too, that we need sophisticated readings to chart how Gissing actively mediates, rather than merely reflecting, the contradictions by which his imagination is compelled.

Rachel Bowlby's *Just Looking: Consumer Culture in Dreiser, Gissing and Zola* (1985) takes up Jameson's themes of naturalist literary doctrine and of the commodification of writing and 'desire', and extends and makes more precise our sense of the context in which Gissing wrote. Bowlby shows that the commercial production and marketing of fiction in the later nineteenth century must be understood in relation to the rapid expansion of consumerism generally. Her two introductory chapters illustrate this development, in the international trade fairs inaugurated by the 1851 Great Exhibition in London and in the spread of department stores where shopping became a spectacle, a 'new bourgeois leisure activity'. Bowlby pays particular attention to the ways in which women were addressed by and implicated in consumer culture – not least in relation to the advertising which it relied on, and which soon became indispensable to the selling of books.

Bowlby attends closely to only two of Gissing's novels, *Eve's Ransom* and *New Grub Street*, but her evidence and analysis are pertinent well beyond these readings. *In the Year of Jubilee*, for instance (which she does not mention), is another text magnetized by the question of how literary 'culture' can – or cannot – be differentiated from the frankly commercial uses for which language is now everywhere mobilized, and of which advertising is the paradigm. Bowlby's reading of *Eve's Ransom*, which highlights how masculine anxieties about status and money seek to assuage themselves in a (failed) bid for heterosexual proprietorship, similarly draws attention to structures of feeling that recur in other works. Bowlby shows that we must read these not just in psycho-biographical terms, but also in terms of how new forms of commerce, work and display in the late-Victorian metropolis were changing the social and imaginary construction of femininity and the lived experience of women. Here she anticipates what has since become one major focus of Gissing criticism.

Bowlby's book exemplifies how literary criticism in the 1980s was beginning to deploy increasingly detailed contextual information; and was becoming, in the process, less clearly distinct from general cultural history. For instance, one of several pertinent illustrations reproduced in *Just Looking* shows the endpapers of the cheap 'yellowback' edition of *New Grub Street* (1891), where a full-page advertisement for baby food faces a page of publicity for novels – including several by Gissing – in Smith, Elder's 'Popular Library'. What remains to be determined, once such contexts are brought home to us, is the extent to which we then read literature as contesting as well as exemplifying the conditions in which it comes to print: the precise question, of course, around which *New Grub Street* circles. Here perhaps Bowlby's reading fails to engage fully with the issues it raises. She concludes her chapter on *New Grub Street* by claiming that the novel shows 'there is nothing outside the rigid dichotomies of value and class entailed by the starting assumption that culture is incompatible with the new commerce and the mass society it caters to' (Bowlby, p.117). This seems a reasonable account of the constellation of writers represented as characters – Reardon, Biffen, Milvain, and the rest. However, it fails to take cognisance of the author of *New Grub Street* (the historical George Gissing, and the authorial position created and implied in the novel), who resists assimilation to either side of the 'dichotomy', both by transcending or evading it structurally and in the fact that this literary engagement with mass culture was among the most commercially successful of Gissing's novels. Here, as in her assimilation of Gissing to a generalized notion of naturalist authorship as normally characterized by 'objective, impersonal language' (p.10), Bowlby makes less than she might of the rhetorical means by which Gissing contrives to disrupt the apparently rigid structures and positions that seem initially to determine his fictions.

John Carey's account of Gissing in Chapter 5 of *The Intellectuals and the Masses* (1992, pp.93-117) is hostile, and reductive. Taking Gissing as the first in a series of 'case studies', Carey finds in his fiction (as in his letters) only what will confirm the pre-established thesis that most members of the 'literary intelligentsia' held anti-democratic views before and during the period of high modernism. It is of course not difficult to isolate such elements: élitist utterances by narrators and characters, or prejudicial characterizations such as that of Jessica Morgan (in *In the Year of Jubilee*), who indeed seems conceived as an illustration of neo-Darwinian theories about the 'intellectual inferiority of women' (Carey, p.99). But only a more dialectical approach than Carey's can make sense of the fact that Gissing characteristically puts élitist and misogynistic views and paradigms to work within narrative contexts that test, and often subvert, their authority.

Carey's book is valuable in encouraging us to read Gissing as shaping the discourse of mass culture (his chapter on 'The Suburbs and the Clerks' is especially useful on this general point) and in locating him as an ideological forerunner of literary modernism. However, its reading of these larger cultural formations is unreflective. Carey characterizes literary modernism one-dimensionally, because he never acknowledges the inherently contradictory aspect of a historical period (the one we still live in) in which a cultural and political democracy was supposedly to be built on the basis of persistent divisions of wealth, labour, and educational opportunity and experience. An adequate reading of Gissing in terms of these historical contradictions must obviously engage with the arguments of the Frankfurt School, dismissed by Carey in a paragraph of mere travesty (p.43), and of later criticism such as Andreas Huyssen's *After the Great Divide: Modernism, Mass Culture, Postmodernism.* Nonetheless Carey's one-sided polemic returns us to a question which plagued Gissing himself and which surely constitutes one reason why his works have continued to interest left-wing critics: the question of whether, and how, intellectuals in conditions of 'mass culture' can sustain a belief in democratic politics while continuing to value complex, or 'élite', cultural works.

The critical engagements we have been reviewing address many broad questions of cultural history – relations between writers and audiences under conditions of 'commodification'; the intellectual's autonomy, or the illusion of it; uncertainties of class status, sought or ascribed, and of political allegiance, in a world where structural economic division coexists with new kinds of social mobility; the implication of masculine sexual desire in anxious and self-assertive responses to new forms and claims of women's subjectivity; the possibility of using the inherited tropes of the novel-form for unaccustomed (unintended?) ends; the legibility of the writer's intention, and the ways the text always exceeds intention. One has only to list these questions to register their continuing (and in some ways renewed) pertinence today. They recur among both the overt themes and the problematic sub-texts of Gissing's work, which makes it doubly rewarding for the kinds of critical attention we have sought to pay it in the present volume.

This collection

These essays extend those critical conversations, exploring Gissing's intriguing ideological ambiguity, his representation of the stresses of modernity, his self-reflexive analysis of the fraught position of the unclassed intellectual, the business of writing and the commodification of culture. But they also broaden them: just as Gissing criticism in the 1970s and 1980s developed the theoretical concerns of those decades, many of the pieces here reflect the wider critical shift towards more historicist, interdisciplinary approaches in recent years by focusing on the extraordinary range of debates that surrounded and shaped his work. At the same time they acknowledge the difficulty of simply placing Gissing's writing, passively, within contemporary discursive frameworks; they recognize his tendency to critically engage with those contexts and transform them, to speak back to the reader in the present as much as the past. And while the essays follow the shifts in Gissing's naturalism: from scenes of urban poverty, to the exile of the lower-middle class intellectual, to the 'vulgarity' of the suburbs, and the whirlpool of the London salon, they trace, too, the underlying preoccupations that transect the different moments of his work.

The first three chapters read Gissing as social investigator and cultural anthropologist. Diana Maltz returns to Gissing's fiction of the 1880s, placing his representation of urban poverty and pauperism beside the debates and practices surrounding organized philanthropy – above all the Charity Organisation Society, which dominated many discussions of urban poverty from the late 1860s by reworking older (Malthusian) debates about the demoralizing effects of charity and poor relief. Gissing himself denied any active involvement with these issues, insisting that his works could play no direct role in social reform; yet he inevitably engaged with the ideological tensions thrown up by contemporary philanthropic movements in the process of developing a realist technique. Gissing represents a shifty criminal underclass playing on middle-class gullibility in the early *Workers in the Dawn* (1880), but if this seems to reinforce the C.O.S.'s stress on the dangers of indiscriminate alms-giving, his representation of charity workers themselves – above all female ones – is far more ambiguous. Maltz traces Gissing's shifting representation of philanthropic women through Gissing's 1880s fiction, from the model philanthropist Helen Norman, who, as a figure from the 1870s, operates as the voice of the C.O.S., to the condescending Miss Lant in *The Nether World*. Contrasting the paternalism of Octavia Hill with the more radical social mission of Toynbee Hall, Maltz concludes with a reading of *Sleeping Fires* (1895), a little-known novel which appears to offer a more positive – if still ambivalent – account of the slum settlement movement, but as a community of female intellectuals as much as a programme of social reform.

Maltz notes Gissing's friendship with Clara Collet as an important source on the Charity Organisation Society. Emma Liggins, too, discusses Gissing's association with Collet as one of a new generation of female social investigators ('New Women' such as Eliza Orme and Clementina Black), and explores the impact of their discussions of various forms of female employment on his own representations of women whose work involved setting themselves, as performers, on public display. Like Maltz, Liggins stresses Gissing's equivocal and ambivalent position as a novelist in contrast with the philanthropic and social debates he draws on. Like the social investigators, he regarded the position of the actress and the violinist – those with artistic aspirations – as being as much shaped by commodity culture as those of the shop-girl or the barmaid. But Gissing's fiction, as well as noting women's vulnerability to exploitation, displays a keen sense of the possibilities open to women in the new kinds of urban spaces that were opening up in the later nineteenth century. Focusing on Clara Hewett in *The Nether World*, Monica Widdowson in *The Odd Women* and Alma Frothingham in *The Whirlpool*, Liggins analyses the correspondences and contrasts between women who occupy otherwise very different class positions, but who all trade on their bodies as an integral component of their work: Alma's position as a 'lady performer' puts her on a level with other sexually dubious figures and so transforms her class position. Gissing thus draws on the research of the social investigators and responds to some of their concerns; but the tension of his fiction depends on the force of these women's ambitions as much as their compromises and failures.

Both Maltz and Liggins chart Gissing's realism through his appropriation and transformation of specific debates surrounding philanthropy and women's work. Scott McCracken, in contrast, maps Gissing's fictional method ethnographically, through his use of that 'most basic and most mediated of cultural forms' – food. McCracken sets out the various positions and spaces that food and eating held within late ninetheenth-century culture – its industrialization and commercialization; its role in blurring the boundary between the public and private spheres; its place within social Darwinist debates on development and sexual selection – as a framework surrounding his reading of *The Nether World*, *Thyrza*, *New Grub Street*, *Born in Exile* and *Eve's Ransom*. He argues that food and eating become complex signifiers in Gissing's fiction precisely because they articulate both a material necessity driven by biology, and an acknowledgement that this is always manifested, performatively, through cultural codes. Emphasizing Gissing's dialogism, McCracken demonstrates how buying and consuming food are always heavily overdetermined acts and spaces of intricate political and symbolic negotiation for Gissing; they also create the possibility of multiple readings and provide an arena for his mordant humour.

McCracken places Gissing's fiction within Darwinian theory as a way of opening up possible meanings rather than narrowing them down. Here he develops the wider interest in the impact of Darwinian and degenerationist theory on contemporary narrative forms that flourished in literary studies during the 1990s, but he also connects them to other forms of social determination – above all, the economy. Gissing's own awareness of these scientific theories and his self-conscious appropriation and adaptation of them has always posed a particular challenge for the modern reader, a challenge which Jenny Bourne Taylor and

David Glover each approach by reading the fiction of the 1890s within the framework of contemporary psychology, while unpacking its wider social significance. Taylor investigates *Born in Exile* – Gissing's most autobiographical and perplexing novel and the one in which the controversies surrounding Darwinian theory most explicitly become part of the protagonist's struggle for social advance and cultural legitimacy – as a case study of a divided consciousness. With *The Private Papers of Henry Ryecroft*, *Born in Exile* is Gissing's most ambitious attempt to represent mental interiority. Whereas the protagonist's unmediated meditations and reflections dominate in *Ryecroft*, in *Born in Exile* there is no identification of narrator and protagonist, as the narrative switches between identification and disengagement. Taylor explores this process, and the social and cultural contradictions that Godwin Peak embodies, through the concept of 'double consciousness' – a term the narrator uses of Peak's own social liminality but which also had a complex but precise psychological significance during the late nineteenth century. Charting the use of 'double consciousness' as a means of investigating memory, consciousness and identity, she discusses how Gissing gives the term itself a double meaning in *Born in Exile* as he reworks the concept within a particular moment in the history of education, while playing on its uses in psychological discourse to destabilize Peak's own identity.

David Glover broadens the focus of Taylor's study. His essay, with those by Simon James and Deborah Parsons, forms a cluster of pieces that explore Gissing's complex response to the stresses of Victorian modernity, particularly in *In the Year of Jubilee* and *The Whirlpool*. The late nineteenth-century discussions of double consciousness, heredity and memory that Gissing develops in his fiction were extended in the 1890s by Sigmund Freud, and while Gissing did not use Freud's work explicitly in his fiction, the writers share strikingly similar concerns during the decade in which psychoanalysis first came of age. Both Gissing and Freud deal with the neuroses triggered by the vertiginousness of modern consumer capitalism, particularly in women; both, too, draw on the dynamic of the accumulation and dissipation of energy, or 'nervous force', that was propounded in the work of Hermann von Helmholtz. Like Taylor, Glover stresses Gissing's use of contemporary physiological psychology – particularly the work of Théodule Ribot – in his representation of the relationship between psyche and soma, and of how the body registers complex mental states within Helmholtz's overall economy of accumulation and dissipation. Glover, however, refers also to the nascent field of psychoanalysis in charting Gissing's symptomatic dissection of the discontinuities and arrhythmia of modernity, as Nancy Lord is swept along in the flow of the Jubilee crowd, Jessica Morgan becomes almost a parody of hysterical decline, and Alma Frothingham treads the fine line between hysteria and neurasthenia. Without claiming any mutual influence, he suggests that Gissing dramatized those social and economic forces that lay behind hysteria but which Freud rarely acknowledged, while Freud would develop his theories of the unconscious and fantasy from cases that echoed the economy of energy and fatigue at work in Gissing's neurasthenic heroines.

Simon James approaches the correspondences between Gissing's and Freud's work from another angle, drawing on Freud's late essay 'Civilization and Its Discontents' to argue that Gissing was concerned to expose the 'pathology' of both masculinity and femininity as a 'disease of civilization'. James notes the conflict between the interiorized self-validation of the intellectual and the pressures of the external world that shapes so many of Gissing's male protagonists, suggesting that, as with other writers of the *fin de siècle*, this leads both to a profoundly uneasy, at times masochistic, identity, and to the splitting of the masculine persona within an identifiable typology. In *The Whirlpool* this splitting is mapped, in the pairing of Harvey Rolfe and Hugh Carnaby, onto the relationship between 'home' and colony, and James explores how the tensions within marriage in *The Whirlpool* are often charted through the opposition between the febrile feminized decadence of the metropolis and the imagined open spaces of the Empire. Unlike Haggard, Gissing does not feminize the colonies; instead they become an unstable, liminal outlet for the energies sublimated by polite society – the place where the self-cancelling drive towards aggression that Freud identified as both the product and the potential destroyer of civilization is at once located and contained.

Deborah Parsons takes up and develops another theme of David Glover's essay. Focusing on Gissing's dramatization of psychic responses to the metropolis, she sets *The Whirlpool* in the context of European naturalism by comparing Gissing's narrative of modernity with those of Emile Zola and Benito Pérez Galdós. While Glover explores how Gissing's appropriation of Helmholtzian science could be used to dramatize the instability of modernity and the mind's response to it, Parsons characterizes late nineteenth-century bourgeois modernity in similar terms – as a heady mix of financial speculation, the profusion of commodity production and the lure of consumption – but argues that the phantasmagoric experience of modern life in London, Paris and Madrid tends ultimately to subvert the 'scientific' naturalism of all three writers. This urban experience, she reminds us, was the context for the emergence not only of naturalism but also of impressionism. Drawing on George Simmel's contemporary analysis of the psychology of the city (both the 1895 essay 'Sociological Aesthetics' and the more famous 'The Metropolis and Urban Life' of 1903), she argues that each writer anticipates modernist narrative in distinct ways by engaging with what Simmel describes as the phantasmagoric consciousness produced by the culture of high capitalism: a consciousness in which the stimuli of over-consumption engender ennui, neurasthenia and disengagement. The novels that Parsons compares illustrate the struggle of realist and naturalist narration to register the unstable world of the new metropolis, and their heroines – Gissing's neurasthenic Alma Rolfe, Zola's Renée Saccard and Galdós's Isidora Rufete – are case studies in its impact on psychology and identity.

The last three essays in the collection all reflect on the conditions and traditions of writing and reading in considering Gissing's role in shaping twentieth-century concepts of the 'unclassed' writer and of the writer as an internal exile. Martin Ryle takes up one of the key concerns of critics – Gissing's sense of the need of the writer to 'keep apart', both politically and aesthetically – and argues that the forms of cultural authority embodied in Gissing's complex performances of disengagement played a specific role in the history of literary modernism by affirming the link between artistic and political sensibility in the very act of apparently denying it. Contrasting Dickens' performance of cultural authority (in which, as Gissing noted, the writer can speak directly to and for the class that constitutes his primary readership) with the 'spirit of antagonism' of the modern writer towards his more fragmented audience, Ryle explores the ways in which the fraught condition of the intellectual is presented through the very texture of the writing as Gissing's writer-protagonists struggle towards a kind of authority and integrity within an increasingly commodified mass culture. *The Unclassed*, *New Grub Street*, *In the Year of Jubilee* and *The Private Papers of Henry Ryecroft* each dramatize aspects of these frustrations in different ways, and show how the writer's inescapable sociality is inscribed in the very process of imagining an escape from society.

Ryle suggests that in their struggle to escape from the literary market, Gissing's writer-protagonists articulate a materialist understanding of culture which would be made explicit in Marxist theory, and this question – of how Gissing seems to anticipate and engage with twentieth-century theoretical concerns – is investigated by Christina Lupton and Tilman Reitz in a close analysis of the nature and effects of self-consciousness in *New Grub Street*. *New Grub Street*, they argue, seems to deny the possibility of a transformative consciousness in which an understanding of the conditions of a situation is a prelude to changing it. Reardon and the other characters finally collude with the cynical opportunism of the victor, Jasper Milvain, by accepting his social Darwinist logic and internalizing its objective force: their understanding of their condition does not prevent their being determined by it. In contrast with the self-conscious narratives of the eighteenth century, *New Grub Street* in its formal self-reflexivity reworks the process of estrangement that the characters experience themselves, and in this respect it anticipates twentieth-century debates in critical theory – above all, Walter Benjamin's concept of reification. Like Gissing, Benjamin was engrossed with the 'hope of pessimism', with the liberating possibilities of representing culture in its most overdetermined form. If we read it in the light of Benjamin's analysis of the shocking, liberating effects of self-estrangement – for readers in the future if not the present – then *New Grub Street*'s self-consciousness may indeed have transformative possibilities.

It is to those future readers that Patrick Parrinder turns in the final essay, as the myth of Gissing's life merges with the voice of the unclassed that he fictionalized to create a presence, a tone as much as a direct influence, that pervades much twentieth-century fiction. Gissing may not have anticipated literary modernism in his narrative technique, but the tensions of social mobility, exile, and lower-middle-class marginality that he embodied and dramatized would persist through twentieth century writing – transformed, sometimes, into comedy or nostalgia. Parrinder shows how Gissing's legendary life-story has been as much appropriated and transmuted as have the characteristic motifs of his writing. Arnold Bennett and Frank Swinnerton both owed Gissing a literary debt while making his failures the subject of their fiction. George Orwell and Jean Rhys each reworked his characteristic guilt and 'double consciousness', which would become an object of satire in Philip Larkin's novels. Gissing's influence also lies in his representation of 'vulgar' uneducated life; and here the proto-modernist text that lies within *New Grub Street*, Harold Biffen's 'Mr Bailey, Grocer', becomes an ironic precursor of late twentieth-century fiction – above all Graham Swift's first novel, *The Sweet Shop Owner*, in which the study of the 'psychology of the suburbs' reaches its height. Parrinder concludes by noting that while we may wish the class-dominated world that shaped Gissing's unclassed heroes was past and gone, the tensions surrounding social mobility and mass education – not to mention consumer capitalism – that his writing grappled with have only grown more powerful. This collection is not the last word on Gissing, but it shows, we hope, how his work speaks to us now, and will continue speaking in the future.

Bibliography

Alden, Patricia (1986), *Social Mobility in the English Bildungsroman: Gissing, Hardy, Bennett and Lawrence*, UMI Research Press, Ann Arbor and London.
Born, Daniel (1996), *The Birth of Liberal Guilt in the English Novel*, University of North Carolina Press, North Carolina.
Bowlby, Rachel (1985), *Just Looking: Consumer Culture in Dreiser, Gissing and Zola*, Methuen, London and New York.
Carey, John (1992), *The Intellectuals and the Masses: Pride and Prejudice among the Literary Intelligentsia, 1880 – 1939*, Faber and Faber, London.
Goode, John (1968), 'Gissing, Morris and English Socialism', *Victorian Studies* XII:2, pp.201-226.
Goode, John (1978), *George Gissing: Ideology and Fiction*, Vision Press, London.
Greenslade, William (1994), *Degeneration, culture and the novel, 1880 – 1940*, Cambridge University Press, Cambridge.
Gross, John (1969), *The Rise and Fall of the Man of Letters: Aspects of English Literary Life since 1800*, Weidenfeld, London.
Grylls, David (1986), *The Paradox of Gissing*, Allen and Unwin, London.
Halperin, John (1982), *Gissing: A Life in Books*, Oxford University Press, Oxford.
Howard, David, Lucas, John and Goode, John (eds.) (1966), *Tradition and tolerance in nineteenth-century fiction: critical essays on some English and American novels*, Routledge and Kegan Paul, London.
Huyssen, Andreas (1986), *After the Great Divide: Modernism, Mass Culture, Postmodernism*, Indiana University Press, Bloomington.
Jameson, Fredric (1981), *The Political Unconscious: Narrative as a Socially Symbolic Act*, Methuen, London.
Keating, P.J. (1989), *The Haunted Study: A social history of the English Novel, 1875 – 1914*, Martin Secker and Warburg, London.
Korg, Jacob (1963), *George Gissing: A Critical Biography*, University of Washington Press, Washington.
Ledger, Sally (1997), *The new woman: fiction and feminism at the fin de siècle*, Manchester University Press, Manchester.
Mattheisen, Paul F., Young, Arthur C. and Coustillas, Pierre (eds.) (1990–7), *The Collected Letters of George Gissing*, Ohio University Press, Athens, Ohio.
Poole, Adrian (1975), *Gissing in Context*, Macmillan, London.
Postmus, Bouwe (2001), *A Garland for Gissing*, Rodopi, Amsterdam.
Selig, Robert L. (1995 (revised ed.)), *George Gissing*, Twayne-Simon and Prentice Hall, New York and London.
Sloan, John (1989), *George Gissing: the cultural challenge*, Macmillan, Basingstoke.
Spiers, John and Coustillas, Pierre (1971), *The Rediscovery of George Gissing*, National Book League, London.
Tindall, Gillian (1974), *The Born Exile: George Gissing*, Temple Smith, London.
Williams, Raymond (1958), *Culture and Society*, Chatto and Windus, London.

Chapter 1

Blatherwicks and Busybodies: Gissing on the Culture of Philanthropic Slumming

Diana Maltz

> The mendicant wore his ordinary garments, for it would have been impossible to find worse, but over each eye he had tied a large green shade, the pair being not unlike the blinkers of horses, which signified that he had sustained the irreparable misfortune of loss of eyesight. He had, moreover, all at once become one-armed, the left being so skilfully disposed that nothing but a close examination could have shown that it was not in reality amputated. On his head was a chimney-pot hat, terribly battered, around which was wrapped a piece of white cardboard, bearing these words, half in written, half in printed, characters:
>
> "CHRISTIEN FRENDS!
> Pray concider a widood Father
> The victim of a Explogion
> And may God bless you."
>
> In his right hand he held a stick, and he directed Arthur to guide him by the empty sleeve on the other side. In this manner they issued out of Whitecross Street and proceeded westwards. (*Workers in the Dawn*, vol. 1, pp.70f.)

The drunken Bill Blatherwick of Gissing's *Workers in the Dawn* (1880) feigns blindness and lameness in order to collect alms from the West End bourgeoisie. One of a shifty criminal underclass, he brings his innate coarseness and sadism to bear in his treatment of the shivering and cowering Arthur, whose blonde angelic looks earn them more pennies from passers-by. Through this scenario, Gissing would seem to affirm the need for institutional checks on almsgiving, and he might be interpreted as approving of the eleven-year-old Society for Organizing Charitable Relief and Repressing Mendacity, or Charity Organisation Society (C.O.S). Established following the bureaucratic disaster of the Mansion House Relief Fund of 1866-9, the C.O.S. acted as a clearinghouse to distinguish between deserving and undeserving recipients of charity.

The scene of Bill Blatherwick and Arthur Golding cadging pennies in the snow not only evokes a Dickensian pathos, but also expresses the doubt of a generation of philanthropists who questioned the effectiveness of almsgiving. Through the end of the century, social reformers and writers remained suspicious of those who appeared penniless but whom they suspected were profiting very well as beggars. In terms of institutional parish support, these reformers sought to track how much each applicant for help was gaining; they were wary that some were benefiting simultaneously from 'overlapping' agencies. Most of all, the middle classes feared pauperization, or the demoralization of the urban working classes. These ideas were articulated in essays and tracts by the founders of the C.O.S: the society's secretary Charles Loch, Helen Dendy Bosanquet and her husband Bernard Bosanquet (whose brother Charles Bosanquet had served as the C.O.S.'s first secretary before Loch), and Octavia Hill.

A dauntless organizer, Octavia Hill systematized visitation into an efficient network that included female settlement house workers, sanitary inspectors, district nurses, and poor law guardians. She further devised the role of lady rent collector as a means of supervising and regulating poor families. In talks that she presented in parish halls, Hill outlined what she called 'a more excellent way of charity', persuading volunteers to eschew random doles in favour of research into each applicant's history and practical aids to the applicant's self-help (Hill, p.63). Hill claimed that small monetary gifts were not merely useless in the long run, but injurious, as they set the poor up to expect further gifts as a matter of habit. Such gifts, she argued, 'eat out their energy and self-reliance' and deprive the poor of self-respect (p.93).

In this essay, I examine Gissing's work in the context of the language of Hill's essays, demonstrating that Gissing was aware of the C.O.S.'s rhetoric and fluctuated in his responses to it. I see this paper as a companion piece to Ruth Livesey's 'Reading for Character: Women Social Reformers and Narratives of the Urban Poor in Late Victorian and Edwardian London' (Livesey, 2004). Livesey argues that lady visitors like Hill's volunteers were taking their cues from realist fiction and gauging the quality of applicants for aid by the conditions of their clothes and homes and contents of their cupboards. In effect, charity volunteers drew a composite picture of each mother and her family by looking for specific physical indications of demoralization or independence. As Livesey writes, 'Character was to be read in the smallest object choices – boots and hats or pawn tickets and rent books – the intimate spaces of the poorest working-class homes yielded up their contents to the reader of character and a moral aetiology was traced into these items' (p.47). Persuaded by Livesey's argument, I would like to augment it by adding that this relationship between caseworkers and novelists was indeed a two-way street. As much as the lady visitors learned from the realist novelists, the novelists also replicated the C.O.S.'s discourse on the able-bodied, undeserving poor. In some novels, Gissing seems to echo Hill's analytic understanding of pauperization, while in others he satirizes her practical efforts. By the 1890s, Gissing joined his contemporaries by looking to simultaneous, parallel models of cross-class relations, models that were as paternalistic toward the poor as Hill's but less emphatic about protecting charities from financial losses.

Gissing had two very real links to the world of Hill and discriminate philanthropy, both of them figures who embodied the active, public New Woman, and both connected to the core network of fifteen or so people at the heart of the C.O.S. Edith Sichel had at the age of 23 joined the Metropolitan Association for the Befriending of Young Servants founded by Janey Nassau Senior. A sister of the novelist Tom Hughes, Senior belonged to the Christian Socialist circle around

F.D. Maurice's Working Men's College with whom Octavia Hill allied herself. Senior's organization worked to save vulnerable women before they fell by assigning runaway girls menial jobs in middle-class homes. Clearly of a philanthropic bent, Sichel had tried in an 1888 article to claim Gissing as a fellow traveller, much to his chagrin (Sichel, 1888). Writing to Sichel, Gissing insisted that he was interested in such questions purely as themes for his fiction. He asserted 'the philanthropic movements of the day are nothing to me save as artistic material' and that 'my books can never be practically useful' (Coustillas, 1987, pp.16f.). Ultimately, even she admitted that his novels were too pessimistic to advocate any programmes for change, for his working-class roughs seemed impervious to improvement.

The same year as Sichel and Gissing's correspondence, the social worker Clara Collet took up residency in the East End and began her work assisting Charles Booth in his monumental survey, *Life and Labour of the People of London*. As an affiliate of the university settlement Toynbee Hall, a District Secretary of the C.O.S., and a participant in various reform-minded clubs, Collet was fully integrated into the philanthropic vanguard, and this familiarity would ground her later critiques of that community. One such critique is Collet's essay, 'The Philosophy of Thrift', published in the *Charity Organization Review* in 1895. Gissing knew of this article as a work in progress, writing to Collet: 'I wonder how you have got on with your uncongenial essay?' (Mattheisen et. al., vol.5, p.282). This is a double-entendre, as the writing of the essay was a burden on Collet, but also 'uncongenial' to her colleagues in the C.O.S. As the editors of Gissing's *Collected Letters* claim, 'Although Clara Collet was connected with the C.O.S. for more than twenty years, she seems to have been uneasy with that institution's emphasis on self-help' (p.283). Gissing himself had no direct business with the C.O.S., and therefore his correspondence with Collet is an important source for our knowledge of how his views on the C.O.S. and its philanthropic project changed over time.

As Deborah McDonald's recent biography of Clara Collet attests, Collet's friendship was extremely significant to Gissing (McDonald, 2004). Their relationship centred on his domestic difficulties and her efforts to relieve them, and Gissing showed little interest in Collet's social investigations. Yet Collet was a conduit through which Gissing could observe clashes in philanthropic theory and practice. She occasionally sent Gissing her published work (Mattheisen et al., vol. 5, pp.118, 140-141; vol.6, pp.4, 60; vol.7, pp.16, 409). Alternately Gissing wrote to Collet for information on social reform groups he wanted to document in his novels such as the anti-smoking lobby and women's journalism schools (vol.8, p.107; vol.5, p.188). It is true that Gissing's questions about Collet's economic writings are brief and most often occur at the end of his letters as a polite gesture. Their correspondence contains little if any developed dialogue about her charitable or sociological work. At the same time, it is worthwhile to recall that many of Gissing's letters were destroyed by Collet.

Like many of his Victorian contemporaries, Gissing found an easy target in the figure of the philanthropic do-gooder. He identified wealthy women for whom charity was a mere star in a constellation of social climbing activities, like Lady Ogram in *Our Friend the Charlatan* (1901) who subsidizes a hospital to thwart her philanthropic rivals, or Sybil Carnaby in *The Whirlpool* (1897) who lectures on the suffering of working girls after demonstrating her maliciousness elsewhere. As John Goode noted in his 1967 analysis of *The Nether World* (1889), Gissing implicitly critiqued the C.O.S and its inherent paternalism through his representation of the arrogant Miss Lant, who doles out judgment more freely than she dispenses cheap soup (*The Nether World*, p.216). A central image of *Will Warburton* (1905) is Norbert Franks' painting, 'The Slummer'. Franks hesitates in determining how he will represent the face of the female district visitor, leaving it for last. In keeping with Franks' (and Gissing's) realist credo, the protagonist advises Franks to depict the girl as ' "sharp-nosed, thin-lipped, rather anaemic, with a universe of self-conceit in the eye" ' (p.62). I mention these instances rather than, say, Mr. Gresham's short-term charity to Arthur Golding in *Workers in the Dawn* or Mr. Tollady's more lasting generosity to him, because in this essay, I am less interested in the upper-class independent philanthropist, or the working person who shows neighbourly kindness to a sufferer in his own class. This essay is more focused on the formation of philanthropic institutions with their own C.O.S.-like principles, and Gissing's representation of their proponents as they toe the party line.

The rhetoric of discriminate almsgiving

Octavia Hill's essays, published in two collections, *Homes of the London Poor* (1875) and *Our Common Land* (1877), clearly capture the spirit and intention of the Charity Organisation Society. They are not dull reading, in part because Hill frequently dramatizes her points through case histories.

> You or I go into a wretched room; we see children dirty and without shoes, a forlorn woman tells us a story of extreme poverty, how her husband can find no work. We think it can do no harm to give the children boots to go to school; we give them and hear no more. Perhaps we go to Scotland the following week, and flatter ourselves if we remember the children that the gift of boots at least was useful. Yet just think what harm that may have done. Perhaps the woman was a drunkard, and pawned the boots at once and drank the money; or perhaps the man was a drunkard neglecting his home, and the needs of it, which should have been the means of recalling him to his duties, he finds partially met by you and me and others; or perhaps the clergy have seen that the poor woman cannot support the children and her husband, who is much too ill to find work, and have felt that if she and they are not to die of starvation they must go into the workhouse, for it is the only means of getting enough for them; charity not being organized in the district, cannot undertake to do all that is wanted for them, and so had better do nothing. For gifts so given may raise false hopes which you and I, now pleasantly enjoying ourselves, never think of. Because we went in and gave those boots, because others like us gave coal-tickets and soup-tickets last winter, what may not turn up? the poor woman asks herself. That gambling, desperate spirit enters into her heart, the stake being freedom and home. She plays high; she wins, or loses.... (Hill, pp.51f.).

Hill argues that instead of saving for a period in which trade is slack, poor families who are accustomed to doles act irresponsibly, failing to send their daughters out for domestic service or to invest in having their sons apprenticed, so that each might be self-reliant. In place of money, Hill and her colleagues try to provide the poor with contacts for jobs, short-term stays outside the city so that the ill may regain their health (to pursue work later on), and savings clubs to encourage habits of thrift. Hill and her fellow-workers also advocated a variety of cultural and aesthetic provisions for the poor's spiritual elevation, such as public libraries, art galleries, open spaces and beautified public thoroughfares (Maltz, 1998).

Hill frequently chastened parishes for the lack of communication between their charitable workers, particularly singling out the species of uneducated but well-meaning lady visitor who has gained no guidance from those with greater knowledge of the parish's poor constituents. The C.O.S. itself existed to close these gaps – for instance, to put ministers and sanitary authorities in contact with voluntary visitors to the poor. Hill critiques the innocent and inadvertently harmful lady visitor who goes out slumming only intermittently and hands out doles to a needy family, no questions asked. She advises that through proper networks of communication, the visitor will be not merely empathetic but informed and far-sighted. Hill advocates knowing the poor not merely as 'poor people', but as individuals through more regular and intimate contact with them (Hill, p.49). The gift that one gives instead of mere money is one's self (p.62).

In *Workers in the Dawn*, Gissing is invested in representing his heroine as a model female philanthropist. Despite her religious scepticism, Helen Norman is in many ways a voice for the C.O.S. She tells her guardian that she will not merely leave it to established charities to do her work, emphasizing the place of personal relations:

> "The efforts of bodies are commendable and excellent – in their proper places. But for the work I see before me, individual effort is alone fitted; of that I am convinced….." (Gissing, *Workers in the Dawn*, vol.1, p.257)

Helen Norman does contribute money, but assisted by a parish minister who knows each case (later, she will associate with the working-class bookseller Mr. Tollady for his knowledge of his neighbours) Helen exists in a fictional universe far simpler than that of 1880s philanthropy, and her dependence on the individuals Tollady and Heatherley obfuscates the complex system of professional bodies (district visitors, health inspectors, social workers) whose tasks Hill negotiated. Helen is a creature of the late 1870s, the period when Gissing was formulating his novel: had he written *Workers in the Dawn* in the late 1880s or early 1890s, he probably would have situated her in the centre of a women's institutional network, like that of Rhoda Nunn and Mary Barfoot. Helen would have been represented as belonging to a movement of female public intellectuals, as has been described by Martha Vicinus, Judith Walkowitz, and Deborah Epstein Nord in their critical work on late-Victorian, middle-class female urban communities (see Nord; Vicinus; Walkowitz).

Quite early in the novel, Helen's mentor, the slum minister Mr. Heatherley, cautions her after she has paid money out of her purse to a young indigent mother: ' "I must warn you, Miss Norman… against being too easily caught by affecting stories" ' (*Workers in the Dawn*, vol.1, p.274). This warning subsequently proves justified. Some chapters later, Helen is deceived by a family who spend her alms on alcohol; she returns to their room to find the furniture she has purchased for them pawned and the whole family, children included, sprawled out drunk (vol.2, pp.23f.). It is a scene reminiscent of Hill's own accounts of the shiftless poor. Once Helen has confessed her mistake to Mr. Heatherley, he replies:

> "I should advise you to be very careful not to waste your money where there appears but slight hope of its doing good. After all, we have but very little power, except where the recipients of our charity come half way to meet us. Happily there are many such instances, and, as a rule, it is not very difficult to discern between honest distress and a true anxiety to take advantage of help." (vol.2, p.24).

In this case, Helen did 'discern' the unworthiness of her charges through the smell of alcohol in the room when she met them; she had just chosen to ignore it.

Workers in the Dawn revisits the problem of the irredeemably coarse and improvident, moving from Arthur Golding's landlady and her gluttonous family, the Pettidunds, to his rescued wife, Carrie Mitchell. Helen Norman's error of giving doles to people who abuse them is replayed in Arthur's sacrifice of his earnings to Carrie, who uses them to purchase alcohol. In contrast, Arthur struggles to economize, and his friend Will Noble heads a savings club for working men. In fact, the idea that doles threaten the independence of the poor is voiced most emphatically by self-educated, skilled labouring men, characters to whom Gissing lends some credibility (vol.1, p.370; vol. 2, p.8).

There are other ways in which the character of Helen Norman can be seen as dramatizing the problems and solutions outlined by Hill. Helen admits that the poor view her as an outsider, and wonders at how she might gain their trust: ' "My speech, my dress, perhaps, revolts them. They think that I do not belong to their class, and, though they take my money, it is with suspicion of my motives" ' (vol.2, p.26). Helen's sense of the gulf between herself and the poor can be read as a confirmation of Hill's views about the distancing effects of giving and receiving doles. Considering this gulf across classes, Helen asks Heatherley: ' "Are they so degraded as to consider themselves hopelessly inferior?" ' – a question which carries the full weight of the C.O.S.'s theories of urban demoralization with it. Helen is dissatisfied by Heatherley's declaration that nothing can be done for the residuum, and asks: ' "Have we no means of raising them?" ' (vol.2, p.24). The two solutions that Heatherley eventually suggests are education and a cross-class trust rooted in Christian practice, both redolent of Hill's own schema. Not surprisingly, Helen goes from merely distributing money to teaching classes to working girls and nursing the ill, and these activities eclipse her almsgiving in the second half of the book.

Helen Norman endures charges that her social work is insincere, and in this regard, Gissing again reprises issues faced by real reformers of the 1880s. These men and women felt the acute need to distinguish their committed slum work from mere social, recreational slumming. (The term 'slumming' actually encompassed both practices: it was defined in 1884 as 'visiting the poor for charitable or philanthropic purposes', but also 'out of curiosity, especially as a fashionable pursuit': see Adburgham, p.143.) As Edith Sichel reflected in 1888, 'Amateur Philanthropy has become as fashionable as Amateur Theatricals have hitherto been' (Coustillas, 1987, p.13). Satirists were discovering plenty of fodder in the image of silly upper-class gentlemen and ladies descending into the abyss for a cheap thrill. In one instance, the cartoonist George Du Maurier posed the slummers in mackintoshes 'to hide [their] diamonds' (*Punch*, 22 December 1883, p.294). Certainly such casual slummers did exist in real life. Gissing picks up on the tension between philanthropic and leisurely slumming when he has Mr. Gresham deride Helen's visits to the slums as mere voyeurism. ' "Let her by all means disgust herself with a peep into these eastern dens of yours," ' he tells the parson (*Workers in the Dawn*, vol.1, p.266). Later this accusation of slumming is more virulently voiced by the working man John Pether:

> "Do you lack amusements? Haven't you your theatres and your balls, your carriages and horses to show yourself with in the park; can't you eat and drink of the best from morning to night? Isn't this enough, but you must look for new excitement in gaols and hospitals and the holes which such as we call homes?' (vol. 1, p. 369).

Helen proves him wrong by returning to the neighbourhood again and again, becoming a fixture there through her regular services to the community. We can spot the philanthropist's eager disavowal of any relation to social slummers in real life, too (see Koven, 2003). During the meeting of the 1884 Royal Commission on Housing, when Octavia Hill was asked about 'the danger... that [lady visitors] are liable to crotchets, and caprices, and fashionable follies', she tersely replied: 'We get rid of the fashionable people pretty soon' (Darley, p. 230).

So, in summary, it would seem that – in 1880, at least – Gissing shared many of the same central policies and anxieties as the C.O.S. But would he remain so affirming? I locate a note of discord in his 1884 novel *The Unclassed*. Here Gissing demonstrates his awareness of trends in institutional philanthropy, for he salutes a programme in tenement reform that directly resembles Octavia Hill's association of 'lady rent collectors':

> Lodgers in the Lane and the Court had come to understand that not even punctual payment of weekly rent was sufficient to guarantee them stability of tenure. Under this singular lady-landlord, something more than that was expected and required, and whilst those who were capable of adjusting themselves to the new regime found, on the whole, that things went vastly better with them, such as could by no means overcome their love of filth, moral and material, troubled themselves little when the notice to quit came, together with a little sum of money to cover the expenses of removal. (*The Unclassed*, p.310)

The main distinction between the real programme and Gissing's fictionalization of it lies in that last line about paying the undeserving to vacate the buildings. No C.O.S. delegate would have sanctioned such a thing, as it was essentially a reward to the unclean and improvident. Gissing is not merely easing the way of his fictional residuum, but also of his reader. The act of eviction is an ugly one, and Gissing appears sensitive to the fact that a popular readership might recoil from any heroine directing it. This detail seems a crack in the plaster that had bonded Gissing to the new philanthropy, a fissure that would widen by the time he conceived of the condescending Miss Lant in *The Nether World* in 1889.

Gissing and the settlement idea

Arguably, Gissing's attitude towards charity evolved along with that of the leading social thinkers in the last two decades of the nineteenth century. Among the original supporters of the C.O.S. were Canon Samuel Barnett and his wife Henrietta Rowland Barnett, the founders of the university settlement Toynbee Hall. The settlement movement grew out of some of the same convictions that grounded Hill's philanthropy, particularly the paternalistic faith in guiding the poor through personal relations (see Briggs; Meacham). However, as much as Hill assented to the benefits of disseminating bourgeois culture to the poor, the Barnetts claimed this as the central tenet of Toynbee Hall's social mission (Koven, 1987). Students and graduates came down from Oxford to live at Toynbee Hall where they gained access to the lives of the poor in East London by living among them as neighbours. Octavia Hill, promoting responsible family life and traditional gender roles, was not an enthusiast of settlement, preferring that female charity workers live outside of the slums and remember their primary roles as daughters and wives (Hill, p.25).

In the mid-1880s, the Barnetts began distancing themselves from the narrowly charitable orientation of the C.O.S. They demonstrated this difference in their defence of unions during the dock strikes of the late 1880s and in their developing support in the 1890s and 1900s for state provisions such as old age pensions and school lunches for poor children. In the context of the changing attitudes and new social forces that would eventually lead to the establishment of the welfare state, Hill's punitive stance towards the undeserving poor seemed old-fashioned and inflexible. The Barnetts were impatient with the C.O.S.'s reluctance to participate in new schemes of reform. In 1895, Samuel Barnett threw down the gauntlet, accusing the C.O.S. of narrowness and dogmatism and claiming it was 'not in sympathy with the forces which are shaping the times' (see Barnett, H., 1919, p.659). In an era of advancing fellow feeling and service, the C.O.S. continued to '[stand] by till people learn their mistakes by common suffering' (ibid.). Barnett hereby adopted what he and his wife would call 'practicable socialism' (Barnett, S., 1894).

While Gissing certainly never grew to adopt state socialism, in the mid-1890s he was showing signs of disenchantment with the C.O.S's ideology. Of course, Gissing had no direct encounters with the C.O.S., but his friend and confidante Clara Collet worked for it. In a letter to Collet in 1894, he expressed a kind of sly delight at her own published satire of charity organization, asking: 'Won't it be thought rather pungent by the C.O.S. people?' (Mattheisen et al, vol. 5, p.191). In February, 1894, Collet had read her essay, entitled 'Three Ideal C.O.S. Secretaries', at the private Denison Club, named for one of the founders of the C.O.S., and it was published in the *Charity Organization Review* the following month. The piece attributed to fictional characters traits all too familiar among charity workers. Through the figure of Antigone, Collet shows how an investment in abstract principles can blind one, robbing one of sympathy. She suggests that like *Middlemarch*'s Dorothea Brooke, the philanthropist may suffer from 'an enlarged conscience' wherein one's eagerness to effect social change is not balanced by recognition of what is right (Coustillas, 1992, p.29). On her fictional

C.O.S. committee, Collet locates Eliot's stuffy Casaubon, gossipy Mrs. Cadwallader and loopy Mr. Brooke as well. Collet demonstrates how this group would squabble in its prescriptions for and against alms, and how each ego would come into play. Mary Garth, whom Collet imagines as a paid secretary, would pacify everyone for the very reason that 'the spirit of the reformer is entirely absent' in her (p.31). Collet claims that in contrast to Mary, women like Dorothea lack a sense of humour and are likely to feel superior to others. Collet ends her essay with a positive account of Shakespeare's Portia, in whom intellect and sympathy are more balanced. All in all, Collet's message about the C.O.S. is an unflattering one. Gissing wrote to Collet that he found the essay 'thoughtful and clear' and 'laughed & admired' it, and this tells us something of his own estimation of the C.O.S. The section that Gissing singles out for special praise is Collet's 'adaptation of *Middlemarch* to modern conditions' — that is, her rendering of the machinations of her imaginary C.O.S. committee (Mattheisen et al., vol.5, p.191).

Published just a year after Gissing praised Clara Collet's playful satire of the C.O.S. and in the same year as the Barnetts' break from it, Gissing's *Sleeping Fires* (1895) ends with a brief recommendation for the settlement idea that had originated with the Barnetts. Here Gissing moves to give the social experiment of settlement greater emphasis over the financial aims of the C.O.S. Gissing had encountered the idea of settlement as early as 1888 when he had read reviews of Mrs. Humphry Ward's *Robert Elsmere*, a novel he criticized because he thought its theme of a protagonist's loss of faith hackneyed (Mattheisen et al., vol.3, p.201). As much as Gissing found fault with this, in his letters he did not critique or even mention *Elsmere*'s advocacy of slum settlement. Such ideas would lie dormant until Gissing wrote *Sleeping Fires* seven years later. (Incidentally, the success of her novel led Mrs. Humphry Ward to found the Passmore Edwards Settlement in Tavistock Square, London.)

Sleeping Fires largely hinges upon the conflict between the inflexible Lady Revill and her ward Louis. Louis has found a mentor in Mrs. Tresilian, a lady of 'advanced opinions' who 'addresses meetings on political and humanitarian subjects' and who inspires him to do 'some work of quiet usefulness' in public life (*Sleeping Fires*, pp.28f., 62). Finding this relationship distasteful, Lady Revill regrets that he 'is going through the stage of hot radicalism' (p.114). When Louis dies of fever before returning from his journey to Greece, the novel ends with Lady Revill's education in philanthropic goodwill. She learns that though 'a good deal of nonsense is talked and believed about her', Mrs. Tresilian is 'a citizeness of the world and lives wherever she can do good' (p.130). Lady Revill repents her conservative prejudices and makes amends by contributing financially to Mrs. Tresilian's settlement in Bermondsey. A mere paragraph is devoted to a description of the settlement project:

> She belonged to an informal sisterhood, who had recently undertaken to live, two or three together, and in turns, among this poor population for example and for help. They kept no servants; all the work of the house was done by their own hands. Each of them took up their abode here for three weeks at a time. (p.167)

In the context of what we know about historic settlements like Toynbee Hall, or more appropriately, women's university settlements, we can see ways in which Gissing is both appropriating their ideas and revising them. To live among the poor 'for example and for help' echoes the Barnetts' dictum that middle-class settlers would civilize their working-class neighbours by example. The claim to 'keep no servants' is an assertion of classlessness that surpasses the Barnetts' standard of democracy: Toynbee Hall did indeed hire help from among the neighbourhood poor and Barnett was insistent that his settlers not imitate the standard of living of the poor (Barnett, H., 1919, p.315).

Yet this detail about housekeeping in Gissing's narrative points to the differences between men's and women's settlements. We have evidence that some of the smaller women's settlements and less formal experiments in working-class neighbourhoods did in fact forego servants. To the amazement of their friend, the poet Amy Levy, Clementina Black and her two sisters Emma and Grace managed their own home on Fitzroy Street in London (Beckman, p.131). So did the trained nurse Honnor Morton and her two friends in opening their settlement in 'a rather gloomy block of artisans' dwellings' in Hoxton (Vicinus, pp.222f.). Gissing's fictional women's settlement resembles Morton's in its small scope. Women's settlements varied in their comparative resources and luxury, and Martha Vicinus observes pointedly that at Morton's settlement, in contrast to the majestic Toynbee Hall, 'money was spent to further the social services, not for the buildings' (p.216).

Mrs. Tresilian's volunteers only live in the slums for stretches of three weeks, a practice which distinguished occasional slummers from more stable settlers, and incidentally, a practice which in real life proved to be a liability for the success of women's settlements. Resigning from the leadership of the Women's University Settlement at the end of its first year, Alice Gruner complained that she had had to assume the unfinished projects of short-term volunteers, most of whom only stayed for three to four weeks (Vicinus, p.224). This anticipated a prevailing tension between long-term professionals and the volunteers whom they accused of dilettantism.

Gissing's choice of a women's settlement for the denouement of his story complicates our analysis in a fruitful way. One would expect that, given his revulsion at vulgarity, he would be especially supportive of programmes for cultural diffusion. Toynbee Hall, for instance, was almost notorious for its alliances with the Aesthetic Movement. (In another essay, I have discussed Gissing's vacillating faith in the practicability of missionary aesthetic schemes: see Maltz, 2000.) In contrast to the men's settlements, women's settlements seem to have been perhaps less ambitiously high-brow in their aims. In the place of lectures on Shakespeare, one finds classes in sewing and cooking or more leisure-minded clubs to get girls off the streets. Female volunteers concentrated on teaching working-class women proper household management, particularly dietetic and hygienic precautions for their children's welfare.

Admittedly, Gissing's representation of the settlement ideal in *Sleeping Fires* lacks precision or detail, and his advocacy is rather half-hearted. Mrs. Tresilian invites the protagonist Langley to befriend her brother: ' "He rails against centralisation, persuades the large people to live at home and be active – and so on.

A good deal of Ruskin in it, of course, but he has ideas of his own" ' (*Sleeping Fires*, p.191). For all that Gissing presents these reformers in a positive light, Langley does not overtly take up the reins of social reform. He does however renounce Greece to return to England. Whether his commitment to his new wife, Lady Revill, will include a share in her philanthropic work is barely implied.

On the basis of the novels I have described here, it is tempting for any critic to claim that Gissing simply rejects the C.O.S's model of discriminate philanthropy and opts for slum settlement instead. If Gissing knew of the widely publicized fissure between the C.O.S. and the Barnetts, we might argue he wrote with specific reference to their disagreement, and claim that he was defying the C.O.S. by consciously endorsing its opponents' own project. We might speculate that, since he had read Collet's own articles in the *Charity Organization Review*, perhaps he had read others: the Barnetts' defection from the fold was the topic of many such essays there in 1895. Yet all this is mere conjecture. As it never comes up in his correspondence with Collet, it is unclear whether Gissing was even acquainted with the breach.

It is impossible for us to establish exactly how much Gissing knew. At all events, he can only be read as disavowing the C.O.S. and advocating settlement schemes if we assume that he was ignorant of the symbiotic relationship between the two. From its inception, the Women's University Settlement was closely linked with charity organization: many female settlers served on local branches of the C.O.S. and advocated its ideology of thrift and self-help (Koven, 1987, pp.390, 392). As a matter of daily praxis, the W.U.S. farmed its middle-class volunteers out as C.O.S. visitors, rent-collectors for Octavia Hill, school managers, and so on (Vicinus, p.231). Indeed, many chose to live in settlements so that they could do their C.O.S. work more efficiently. Despite her reservations about female settlement, Octavia Hill was the W.U.S.'s first choice as a replacement when the first head Alice Gruner resigned, and she appears to have considered the offer seriously before rejecting it. Ultimately, Hill agreed to sit on the Executive Committee of the W.U.S. in Southwark (ironically, District Secretary and sometime detractor of the C.O.S. Clara Collet sat on the same Executive Committee where she represented the University of London: see Koven, 1987, pp.366ff., 376). Given the fact that Collet sent many of her writings to Gissing and that he read them, Gissing would likely have known of the interdependency between W.U.S. and C.O.S. Yet even this knowledge is not evidenced through their letters.

In representing female settlement in *Sleeping Fires*, Gissing need not be seen as rejecting the discriminate philanthropy of the C.O.S., but he is clearly shifting his interest from the narrow question of money- and alms-giving to broader social issues. By the mid-1890s, the C.O.S.'s exhortation against alms must have indeed sounded like an old tune replayed over and over. It was time for this novelist to discover new creative ground.

One can see how the idea of settlement would have appealed to Gissing. The women who had sought to live in settlements were in fact pursuing the intimate communal life that Gissing envied of university men – only in impoverished districts. Preferring a Ryecroftian escapism into the country, Gissing had no desire to retrace his own steps into East End misery. But he would have liked the idea of a coherent programme that would reform the denizens of the nether world, including the foul-mouthed landladies of his own unfortunate acquaintance.

Like his attitudes towards late-Victorian feminism and socialism, Gissing's responses to institutional philanthropy were complex and inconsistent. I began this essay by listing Gissing's satirical, bitter representations of charity women. In fairness, one must offset these with the Helen Normans and Mrs. Tresilians of his work. Lady Revill's conversion to social reform near the end of *Sleeping Fires* in fact resolves some difficulties posed at the conclusion of his more autobiographical novel *Born in Exile* (1892). There, just as we meet a woman pursuing a satisfying public life as a writer and humanitarian in the East End, we hear another character admit that her 'elegant leisure' is in fact a 'wearisome idleness' that she will have to endure for years (*Born in Exile*, pp.464, 466). Described in these terms, her future makes philanthropy look good by comparison. Through the figure of Lady Revill, the listless aristocrat gets a second chance, choosing activity and meaning over dullness and fashion. Revill's husband Langley likewise resolves to 'shake off this sorcery of Athens,' because, 'living in the past, [one] forgets the present altogether,' and after all, his modern, more mindful life is 'now to begin' (*Sleeping Fires*, p.210).

Bibliography

Adburgham, Alison (1961), *A Punch History of Manners and Modes, 1841-1940*, Hutchinson, London.

Barnett, Henrietta Rowland (1919), *Canon Barnett: His Life, Work and Friends*, J. Murray, London.

Barnett, Samuel, and Henrietta Barnett (1894), *Practicable Socialism; essays on social reform*, Longmans, Green and Co., London and New York.

Beckman, Linda Hunt (2000), *Amy Levy: Her Life and Letters*, Ohio University Press, Athens.

Briggs, Asa (1984), *Toynbee Hall, the First Hundred Years*, Routledge & K. Paul, London, Boston.

Collet, Clara (1895), 'The Philosophy of Thrift', *Charity Organization Review*, October, 410-419.

Coustillas, Pierre (1987), *Brief Interlude: The Letters of George Gissing to Edith Sichel*, Tragara, Edinburgh.

Coustillas, Pierre (1979), 'Gissing's Correspondence with Clara Collet, M.A., Social Investigator and Reformer', *Gissing Journal* XV, 1, 27-34.

Coustillas, Pierre (1992) ' "A Great Deal of Brain to the Square Inch": A Forgotten Essay by Clara Collet', *Gissing Journal* XXVIII, 2, 18-33. Originally in Collet, Clara (1894), 'Three Ideal C.O.S. Secretaries', *Charity Organization Review*, February, 119-28.

Darley, Gillian (1990), *Octavia Hill, a Life*, Constable, London.

Du Maurier, George (1883), 'Overdoing It', *Punch*. December 22, 294.

Gissing, George (1985; first publ. 1892), *Born in Exile*, Hogarth Press, London.

Gissing, George (1907; first publ. 1889), *The Nether World*, Smith, Elder, London.

Gissing, George (1896; first publ. 1895), *Sleeping Fires*, Appleton and Co., New York.

Gissing, George (1968; first publ. 1884), *The Unclassed*, AMS Press, New York.

Gissing, George (1981; first publ. 1905), *Will Warburton*, Harvester Press, Brighton, Sussex.

Gissing, George (1985; first publ. 1880), *Workers in the Dawn*, Harvester Press, Brighton, Sussex.

Goode, John (1967), 'George Gissing's *The Nether World*', in *Tradition and Tolerance in Nineteenth-Century Fiction; Critical Essays on Some English and American Novels*, David Howard, John Lucas and John Goode (eds.), Barnes and Noble, New York.

Hill, Octavia (1877), *Our Common Land (and other short essays)*, Macmillan, London.

Koven, Seth (1987), *Culture and Poverty: The London Settlement House Movement 1870-1914*, Ph.D. thesis, Harvard University.

Koven, Seth (2003), *Slumming: Eros and Altruism in Victorian London*, Princeton UP, Princeton.

Livesey, Ruth (2004), 'Reading for Character: Women Social Reformers and Narratives of the Urban Poor in Late Victorian and Edwardian London', *Journal of Victorian Culture* IX, 1, 43-67.

McDonald, Deborah (2004), *Clara Collet 1860 – 1948: An Educated Working Woman*, Woburn Press, London and Portland (OR).

Maltz, Diana (1998), 'Beauty at Home or Not? Octavia Hill and the Aesthetics of Tenement Reform', in *Homes and Homelessness in the Victorian Imagination*, Murray Baumgarten and H.M. Daleski (eds.), AMS Press, New York, 187-211.

Maltz, Diana (2000), 'Practical Aesthetics and Decadent Rationale in George Gissing', *Victorian Literature and Culture* XXVIII, 1, 55-71.

Mattheisen, Paul F., Young, Arthur C. and Coustillas, Pierre (eds.) (1990-1997), *The Collected Letters of George Gissing* (in nine volumes), Ohio University Press, Athens (Ohio).

Meacham, Standish (1987), *Toynbee Hall and Social Reform 1880-1914: The Search for Community*, Yale University Press, New Haven.

Nord, Deborah Epstein (1995), *Walking the Victorian Streets: Women, Representation, and the City*, Cornell University Press, Ithaca.

Sichel, Edith (1888), 'Two Philanthropic Novelists: Mr. Walter Besant and Mr. George Gissing', *Murray's Magazine* III, 506-18.

Vicinus, Martha (1985), *Independent Women: Work and Community for Single Women, 1850-1920*, University of Chicago Press, Chicago.

Walkowitz, Judith (1992), *City of Dreadful Delight: Narratives of Sexual Danger in Late-Victorian London*, University of Chicago Press, Chicago.

Chapter 2

Her Appearance in Public: Sexual Danger, Urban Space and the Working Woman

Emma Liggins

In the first volume of Gissing's *The Whirlpool* (1897), when the middle-class heroine Alma Frothingham is pursuing an independent lifestyle in Germany in the hope of becoming a professional violinist, she is horrified at the proposal of the millionaire Cyrus Redgrave that she accompany him to his Italian villa:

> She was pale and profoundly agitated. So this was his meaning? – made plain enough at last, though with the most graceful phrasing... Was she not a lady? And who had ever dared to offer a lady an insult such as this? Shop-girls, minor actresses, the inferior sort of governess, must naturally be on their guard; their insecurity was traditional; novel and drama represented their moral vicissitudes. But a lady, who had lived in a great house with many servants, who had founded an Amateur Quartet Society, the hem of whose garment had never been touched with irreverent finger – could *she* stand in peril of such indignity? Not till now had she called to mind the forewarnings of Sibyl Carnaby, which, at the time of hearing them, she did not at all understand. "People," said Sybil, "would approach her with strange ideas." This she might have applied to the grotesque proposal (as it seemed to her) of Felix Dymes, or to the risk of being tempted into premature publicity by a business offer from some not very respectable impresario. What Sibyl meant was now only too clear.... She durst not leave the house for fear of encountering Redgrave, who, if they crossed by chance, might fancy she invited another meeting. (*Whirlpool*, pp.85f.)

Alma's shocked reaction condenses a number of key issues about the sexual behaviour of the working woman and the kinds of 'grotesque proposals' and illicit meetings she invites by entering the public sphere. As Judith Walkowitz has argued, 'entering public space placed women of all classes, whether shop girls or shopping ladies, in a vulnerable position' (Walkowitz, 1992, p.46): outside the safety of the private sphere they ran the risk of compromising their respectability by interacting with unknown men. By occupying the city for the purposes of work and leisure, the young woman may be 'approach[ed] with strange ideas' or 'stand in peril of indignity' if she does not heed the warnings: the chance encounter which Alma fears was one of the dangers of the metropolitan lifestyle, according to an 1892 article by Evelyn March Phillipps on the 'working lady in London' (Phillipps, p.200). What interests me here is the set of assumptions invoked about the sexual behaviour of women in occupations which require them to display themselves in public. Although as a lady Alma seeks to distance herself in class terms from shop-girls and actresses, as an aspiring professional musician she places herself on their level sexually: like the actress, she has to face 'the risk of being tempted into premature publicity' by unscrupulous agents, and she is as liable as the shop-girl to be commodified and 'insulted' by men. Gissing's knowing reference to the 'moral vicissitudes' of shop-girls and actresses who need to 'guard' themselves against unwanted male attention relies on accepted views of their uncontrolled sexual availability and entrenched 'judgements about [their] sexual morality and conduct' (Davis, p.xiv).

Gissing's intervention in debates about women's work came at a 'timely moment', when 'the public mind had become agitated upon the subject of women's labour' (Bulley, p.39). From the 1880s onwards a new generation of female social investigators, including Eliza Orme, Clara Collet, and Clementina Black, an active member of the expanding trade union movement, began to collect statistics and interview women about their lifestyles and working conditions (Levine, pp.109ff.). An 1892 government report on the conditions of women's industrial work in Britain, to which Orme and Collet contributed, was typical in its aim of investigating not only women's wages but also their 'alleged grievances' and 'the effects of employment upon health, morality and the home' (Bulley, p.39). This concern with the morality of female employees encompassed cultural fears about the respectability and reputations of working women as well as the unwomanliness they risked by entering the labour market as competitors with men. Deborah Epstein Nord has highlighted the female social investigators' interest in the working-class mother, suggesting that they 'directed their critical energies towards bolstering women's domestic role' (Nord, p.207), but this fails to take into account their observations of other issues such as women's mobility in the city, their living arrangements and their sexual behaviour. M.M. Bird's later study, *Woman at Work*, underlined the threat of sexual corruption in the workplace and directed women towards 'safe and regular occupation' (Bird, p.11). Discussing acting, she warned:

> In this section of woman's work we have once more returned to that undesirable position from the point of view of the earnest worker, where personal appearance is an important factor. Beauty of face and figure, a pretty voice, charming ways, and personal magnetism have their solid market value here. And where this is the case, the work has always a rather deteriorating effect upon the woman. (pp.240f.)

Women entering certain professions 'with nothing more solid and serviceable than a pretty face' (p.238) are seen to be susceptible to the 'deteriorating effect' of entering the public sphere on men's terms, an 'undesirable position' for the 'earnest worker' the social investigator often sought to address.

Such fears about the effects of work anchored in a woman's 'market value' helped to establish a set of links between women in 'unsafe' professions where their virtue was at risk. Bird goes on to compare the actress to the shop-girl, in order to express her disapproval of occupations where 'personal appearance is found so valuable to the worker' (p.241). Concern about the sexual morality of women workers was harnessed to a prevailing uncertainty about their class identities, particularly in new occupations for women. For example, an 1887 report identifying five classes of women workers included shop-girls and book-binders under 'industrial' and teachers and actresses under 'professional' (Robinson, p.51), whereas Clara Collet's alternative interpretation of 1881 census returns revealed that the largest group in London comprised servants, dress-makers, milliners, teachers, shop-assistants, clerks and bar-maids, most of whom were 'drawn from the lower middle class, and a few from the professional class' (Collet, p.545). Fears about the loss of respectable femininity for the woman worker betrayed widespread anxieties about both class and gender, at a time when educated women's increased entry into the labour market threatened class boundaries.

Gissing's views on women's work have largely been considered in relation to his realism, his participation in debates around educated women's entry into the new professions, and, more recently, in relation to theoretical readings of female urbanism. Early reviewers drew attention to the accuracy of his depiction of women's work cultures. Gissing first became aware of Collet in 1893 when she spoke admiringly of his representations of the urban poor in a lecture for the Charity Organisation Society (some aspects of their relationship are discussed in Diana Maltz's chapter here). His research for his fiction increasingly relied on the consultation of contemporary sources, such as the *Englishwoman's Year Book*, for up-to-date information about wage-levels, hours and working conditions. In 1894 he recommended a recent manual of female employment to his novelist brother Algernon (Mattheisen et al, 1994, p.188) – and his letters regularly include requests for details about employment patterns, salaries and professional training. His protests against the long hours, insanitary conditions and limited leisure time of working women borrowed from social investigatory materials such as the Board of Trade reports on female employment that Collet sent him (Mattheisen et al., 1995, p.4). However, his early depictions of the underpaid and exploited London work-girl have attracted less attention than the 1890s fiction, when he began to focus on women entering the new professions, such as clerical work and journalism, and to consider the disadvantages and freedoms of the educated woman worker. In 1986 David Grylls argued that 'despite his carefully researched information, Gissing's conception of female employment was largely conservative', as his reservations about the effects of education seem to ensure that his female characters 'either do not have a job at all, or have a traditionally "feminine" job' (Grylls, p.159). This view has now been challenged: David Kramer has acknowledged Gissing's recognition of the ambiguities of women's new lifestyles, while Maria Teresa Chialant has argued that he 'introduced "the woman in public" under different guises' – relating both working women and middle-class heroines to 'the new roles played by women in the city at the end of the nineteenth century' (Kramer; Chialant, pp.55, 58). This essay builds on their perceptions.

Gissing's preoccupation with the public woman was established in his first novel, *Workers in the Dawn* (1880), where the prostitute's urban confidence and mobility in the streets invite censure from men unable to accept a woman's ease of movement in the public sphere. However, the figure of the prostitute, 'the quintessential female figure of the urban scene' (Walkowitz, 1992, p.21), also lurks behind his representations of other public women who trade on their physical attractions in the workplace, such as the actress, the shop-girl and the professional musician. On one level, the narratives play out the fears of the social investigators about women's vulnerabilities in professions where the need to 'make a good appearance' (Morley, p.303) was part of the job specification, showing how women's economic exploitation and the working conditions they are forced to accept can intensify the sexual dangers of their occupations. In *The Odd Women* (1893), the shop-girl is not endangered in the work-place but by adopting the life of sexual freedom encouraged by her employers. But in *The Nether World* (1889) and *The Whirlpool* (1897), which I will concentrate on here, Gissing's attack on a system that makes women reliant on men to aid their entry into the public sphere is much more ambiguous, flawed by a tendency to punish women for compromising themselves by their choice of profession. Despite their class differences, the bar-maid turned actress Clara Hewett and the violinist Alma Rolfe encounter similar difficulties in pursuing their careers, as if Gissing is endorsing anti-feminist commentaries on the dangers of the performing woman's public life. A narrative paradigm is established in which the heroine compromises her reputation by her choice of a profession that relies on her sexual attractions, and then has to be subjected to restrictions on her mobility in the form of a suburban marriage. I want to consider the workings of this paradigm in relation to feminist interpretations of the 'public woman' and what Maria-Teresa Chialant describes as 'the "liminal status" occupied by those women who entered public space in the 1880s and 1890s' (Chialant, p.65). Entry into the performing professions, or the act of serving men as customers, necessitate sexual and financial exchanges with men which put women on a level with the prostitute, as the texts show how quickly they become indebted to the men they encounter in the public sphere.

The public woman and her sexual availability

Women who put themselves on public display for a living at the end of the nineteenth century clearly jeopardised their respectability by attracting comparisons with the prostitute, not least in their occupation of urban space. Shop-assistants, bar-maids and actresses often worked until late in disreputable areas. The *Girl's Own Paper* complained in 1896 that bar-maids might be obliged to walk home 'after midnight through some of the worst streets in London' when trains and omnibuses had stopped running (Anon., p.330), symbolically aligning them with the streetwalker. Walkowitz classifies the shop-girl as one of the new social actors liable to be mistaken for the prostitute in the commercial districts of London's West End (1992, p.50; 1998, p.17). The proximity of theatres and concert-halls to the West End haunts of prostitutes also reinforced the enduring popular association between actresses and prostitutes, both seen as 'self-sufficient women who plied their trades in public places' (Davis, p.100). Such associations permeate late

Victorian fiction, in which prostitutes have often previously worked as actresses or shop-girls, as for example in the striking opening of Emile Zola's *Nana* (1880), where the eponymous courtesan heroine appears almost naked on stage in a theatre crowded with men transfixed by her voluptuous body. A 1914 assessment of the profession highlighted the 'humiliating and degrading philandering, a mauling sensuality which is more degrading than any violent abduction' with which actresses had to contend (Morley, p.307), a view of the actress explicitly endorsed by Gissing when he comments in *The Nether World* that Clara Hewett has 'sold herself to degradation' (*Nether World*, p.207). Clara's relationship with Scawthorne, who lures her on with promises of contacts in the theatre world, encourages her to behave like a prostitute: in a letter demanding a push up the career ladder, she draws the line at doing ' "something disgraceful" ', but confesses ' "for a large sum of money there are few things I wouldn't do" ' (p.200).

Being treated as a sexual object was only to be expected given the set of assumptions about women in these professions. Both the 'pretty shop-girl' operating solely 'for the service of the public' (Bird, pp.69, 75) and the bar-maid who 'acts as a decoy, [and] attracts customers' (p.85) have to accept their dependence on male appreciation of their charms. Like the shop-girl behind her counter, the bar-maid positioned herself in front of the commodities, 'suggesting that she herself might be an article for purchase or consumption' (Bailey, p.151). Paula Gillett notes the opposition to musical careers on the part of middle-class parents because of 'the belief that women compromised respectability by making themselves objects of the public – that is, of the male – gaze' (Gillett, p.7). The heroine's chosen career as a professional violinist in *The Whirlpool* is seen as rather scandalous by her step-mother: 'Did Alma forget that her appearance in public would give occasion to most disagreeable forms of gossip?' (*Whirlpool*, p.248). The scene of her first recital makes her into a sexual object, lingering on 'the beautiful curve of her lips' and 'wonderful white shoulders' (pp.33, 36), emphasizing what Gillett refers to as 'the sexual dimension of female musical performance' (Gillett, p.11). Despite a growing recognition of the legitimacy of stage careers for ladies (Jeune, 1893, p.5), stigma still attached to women who worked in the theatre (Davis, p.49). Women in these professions who had to capitalize on 'the advantage of good looks' mobilized fears around sexual respectability (Bird, p.85).

Women entering professions linked to sales or performance relied on their physical appearance and their ability to attract men. A photograph, or the popular *carte de visite*, became a requisite for job applications by bar-maids (Bird, p.85), and advertisements of forthcoming performances often reproduced visual images of actresses and musical women. This was also linked to age. According to Bird, the service industry only employed women under 35, with some landlords preferring bar-maids with 'no previous experience' whose 'innocence' would be part of their appeal (Bird, pp.84f.), and Davis has noted that 'like shop girls and bar-maids [actresses] were virtually unemployable after their looks faded' (Davis, p.48). In *The Nether World*, Clara's mother rightly suspects that she has only been employed ' "because she's good-lookin', an' she'll bring more people to the bar" '

(*Nether World*, p.18). In Theodore Dreiser's slightly later American novel *Sister Carrie* (1900), the heroine's attempts to find work in either the department stores or the theatres in Chicago are all met with appraisals of her physical attractions: her 'trim appearance' invites 'a little flirtation' and the offer of lunch from one theatre manager, and in the stores she shies away from the 'disagreeable familiarity' of a manager who 'ogled her most salaciously' (Dreiser, pp.256, 259). As an actress, the 'sensually attractive' Clara is always described in terms of her 'womanly outline' and 'the dark abundance of her tresses' (*Nether World*, p.282), which she learns to dress for maximum impact on men. The dominant cultural view of the 'lady violinist' as an alluring figure (Gillett, p.113) is reinforced in Mrs Strangeways' flattering comment to Alma that ' "one has to take into account other things besides mastery of the instrument; with the public naturally, a beautiful face and a perfect figure..." ' (*Whirlpool*, p.199). Gissing's uneasy fascination with women who display their bodies is also apparent in the description of Clara's new position as bar-maid:

> Through the day and through the evening Clara Hewett took her place behind Mrs Tubbs's bar. For daylight wear, the dress which had formerly been her best was deemed sufficient; it was simple, but not badly made, and became her figure .... When Clara had surveyed [her evening attire] in the bar mirror, she turned away contemptuously; the material was cheap, the mode vulgar. It must be borne with for the present, like other indignities which she found to be inseparable from her position. (*Nether World*, pp.78f.)

The mirrors behind public bars and the clothes required for bar-work, rather like the low-cut dresses musical women were encouraged to wear (Gillett, p.113), emphasize the female figure in a cheap and vulgar manner. In trades where 'appearance may mean her livelihood as a worker', clothes 'assume a horrible importance' in the drive to attract men (Morley, p.304). Clara's unfeminine self-awareness of her attractions – she considers herself 'as good-looking' as the photographed actresses displayed in a nearby shop-window (*Nether World*, p.82) – seems to invite criticism, as women who are able to bear the 'indignities' of their position in the public eye must also accept that they, like the prostitute, have 'something to sell' (Morley, p.299).

Woman's appearance in public, like her newly acquired shopping habits, fully embeds her in the burgeoning commercial culture, promoting what Walkowitz refers to as 'a marketable, sexualised femininity' (1998, p.7). Clara eagerly anticipates being 'known as a successful actress' so that her portrait will appear 'in the windows' (*Nether World*, p.82), but for Gissing this visual display of the female body, like the photographs of Alma appearing in the press, is seen to be inappropriate publicity, cheapening women's artistic success. The advertisement of performances requiring the printing of women's names on bills and in cheap newspapers reinforces the grossly commercial aspects of their work. Harvey Rolfe experiences 'a sense of degradation' at the sight of his wife's name 'exhibited in staring letters' outside the concert hall and is 'disgusted' by her press announcements (*Whirlpool*, p.308). In her account of the actress's attempts to 'sell' herself, Lena Ashwell highlights her obligation to 'put on a desperately photographic pose and expression, in case the agent's eyes light on [her]' (Morley, p.301), indicating that the performing woman must constantly objectify herself in order to impress men.

This desperate need for display is harnessed to the vulgarity by which Gissing is both fascinated and repelled, as indicated when Rolfe reacts to Alma's playing by feeling that the violinist 'absorbed the vulgarity of her atmosphere. All she did was for effect' (*Whirlpool*, p.41). Simon James has highlighted the commodification of female sexuality and metaphors of prostitution in the later novel, seeing Alma's 'love of display' as part of the process by which she becomes 'another spectacular commodity on offer in the metropolis' (James, p.18). However, his claim that Alma does not sell her body to men in the same way as Sibyl Carnaby (whom we suspect of sleeping with Redgrave to guarantee his investment in her husband's company) needs qualifying. While Gissing may appear to be dramatizing the warnings of the social investigators about women having to put themselves on the market, it remains unclear whether women have the capacity to resist their commodification in commercial culture or have no choice but to 'succumb to spectacle and publicity' (James, p. 20).

The sexual dangers of the workplace and the city

Women who stand behind a bar or a counter, or perform on a stage, were widely perceived as advertising their sexual availability, and had to learn how to deal with unsolicited male attention. Bars were seen as particularly contaminating to female virtue. *A Dictionary of Employments Open to Women*, published by the Women's Institute in 1898, noted the agitation against the employment of women in bars and public houses (including Eliza Orme's 1893 study) because of its 'demoralising tendencies', but concluded that many workers had indicated it 'would be injurious to women's interests to withdraw any opportunities of gaining an honest livelihood' (Philipps, p.16). Although the landlady Mrs Tubbs is not a sexually threatening figure, as an employer she still controls her bar-maid's reputation: after Clara pre-empts her own dismissal by leaving, Sidney reports that the landlady is 'doing her best to take away your good name' (*Nether World*, p.93). The necessity for the actress to learn both to 'fend off unwelcome sexual advances' and to 'constantly sue for men's attention' (Davis, pp.7, 100) can equally apply to women in other professions organized around male-controlled workplaces.

Gissing's women often seek to defy predatory men by carefully modified speech. An experienced bar-maid interviewed by Bird pointed out that ' "a woman's position here is much what she herself makes it. If she refuses to submit to bad language and disrespect, the landlord will quickly stop it" ', though another interviewee appeared less in control, having been ' "turned out into the streets at midnight" ' because she had repulsed the advances of her employer (Bird, p.86). As a bar-maid Clara perfects a 'customary cold reserve' (*Nether World*, p.85) for dealing with her customers and later adopts a tone of 'insolence' with the theatre manager, which helps to ensure that he does no more than 'scrutinise' her. In an attempt to counter Dymes' 'free talk', Alma corrects him for his over-familiarity in using her first name, but she finds it more difficult to defend herself against the 'insolence' and 'impertinence' more stereotypically associated with male speech. Despite her 'severe' request that Dymes not discuss her with a salesman in the music shop where they meet, once she leaves he manages 'with leers and half-phrases' to suggest that she finds him 'fatally attractive' (*Whirlpool*, p.196). Walkowitz argues that in order to combat street harassment, women's magazines instructed ladies 'how to navigate and manage heterosexual relations on the street', encouraging the respectable woman to develop strategies 'to show that she was not available prey' (1998, p.7). However Gissing suggests that dealing with sexual harassment in the workplace requires a woman not only to be constantly on her guard, but to actively defend her right to work unmolested.

It is significant that the performing woman finds herself in debt to the men without whom she would be unable to achieve her professional ambitions. In her list of the skills women had to develop in working in the theatre, Davis includes 'self-sufficiency, mobility, and the freedom to interact with men as colleagues, admirers, pursuers, and economic equals' (Davis, p.16). But Gissing's fiction tends to suggest that the freedom to interact with men inevitably emphasizes women's sexual vulnerabilities and their financial dependency, reinforcing their associations with the prostitute. This is particularly significant in relation to women pursuing careers in the arts, who must rely on male agents for publicity and contacts. In *The Nether World* Gissing elliptically comments on 'various inconsistencies in [Clara's] behaviour' to Scawthorne who claims to have 'secured [her] the opportunities that are so hard to find' (p.273); and the feeling that she remains 'in his debt' suggests some kind of sexual exchange. The theatre manager, who already views her as 'an object of interest', then advances her money to buy clothes for her new role with the unspoken proviso that she will repay him with her body.

These unspoken provisos proliferate in *The Whirlpool*, as making the transition from amateur to professional violinist puts Alma in 'a fatal indebtedness' (p.310) to both Redgrave and Dymes. Paula Gillett has argued that Alma is one of a number of musical heroines who are 'portrayed wholly without sympathy as women who cynically exploit male susceptibility to the charms of the female violinist to their own illegitimate ends' (Gillett, p.130), but surely the narrative is intent on proving that Alma is much more exploited than exploiter. Although she congratulates herself on keeping both men at a distance whilst benefiting from their financial support, she 'distrusted her power to stand alone' (*Whirlpool*, p.261), revealing the performing woman's enforced reliance on sexually predatory men. The sinister Redgrave has a powerful hold over her: she recognizes that he 'had drawn his nets about her; at each meeting with him she felt more perilously compromised ... but she was brought to the point of regarding him as inseparable from her ambitious hopes' (p.261). Her uneasiness about accepting Dymes' offer to pay her advertising expenses prompts her to wonder whether it may have 'occurred to his outrageous vanity that she might be persuaded to become his debtor, with implied obligation of gratitude?' (p.239). The sleazy composer always suggests that they go to his rooms when they meet to discuss matters of business, and there is an unpleasant scene when he tries to force his attentions on her in a hansom cab. The female performer then cannot escape being 'perilously compromised' in the encounters with men necessitated by her choice of career.

The tendency to see the bar-maid as particularly vulnerable is reinforced in both social investigatory accounts and fiction. A rather alarmist article in the *Girls' Own Paper* in 1896 asserted that 'of all occupations undertaken by girls for a living, there is none more difficult and dangerous than that of serving at a "bar", or waiting in restaurants', commenting particularly on the 'language, jests and oaths' of men and the temptations of drink, prostitution and even suicide which these 'weary, lonely girls' must bear (Anon., pp.329f.). A pretty Pimlico bar-maid interviewed by J.T. Grein in the late Victorian period complains about the 'liberties' taken by her customers and the dangers of attracting the governor, who ' "leaves you no peace; he either plagues you or bullies you, and all because you want to be respectable" ' (Grein, p.37). In Thomas Hardy's *Jude the Obscure* (1895) the sexual immorality of the voluptuous Arabella Donn is presumed to be exacerbated by her choice of bar-work: 'She may be getting [coarse] by this time under the duties of the public-house' (pp.327f.). Clara finds the behaviour of the 'vulgarly familiar' and 'insolent' male customers 'hateful beyond anything she could have conceived' (*Nether World*, p.84). Despite arguing for the merits of bar-work over factory employment, her would-be sweetheart Sidney Kirkwood later sneers: ' "I hope you'll enjoy the pleasant, ladylike work you've found! I should think it'll improve your self-respect to wait on the gentlemen of Upper Street!" ' (p.32), suggesting that because she passively accepts male attention the bar-maid has no claim to respectability.

In Gissing's fiction, lower-middle-class men generally take it for granted that bar-maids will quickly agree to offers of marriage and make submissive wives, again denying them any sexual agency of their own. In *The Odd Women* (1893), Edmund Widdowson's brief conversation with an unnamed bar-maid prompts the thought that his ideal wife is much closer to the socially and intellectually inferior woman who serves him than the more independently thinking Monica. The bar-maid is similar to the 'neat, decent and quiet' figure of Grein's description. Widdowson sees her as 'a young woman of some charms, and what appeared to be unaffected modesty', 'a good, nice sort of girl' (*Odd Women*, p.238). His thoughts reveal the contradictions of the male customer's attitudes to the bar-maid: he laments her position behind the counter 'hearing doubtful talk and seeing very often vile sights', but also fantasizes about her docility as a potential wife, remembering her face 'with a complacency which soothed his feelings' (p.238). The complacency with which men treat the bar-maid, despite her apparent modesty, appears to bear out Peter Bailey's comment on the 'male perception at play, half fantasy and half calculation – that of the bar-maid as potential mistress or "kept woman" ' (Bailey, p.163). Women's attempts to resist their categorization as available are shown to have limited success in the texts, suggesting their inability to manipulate the parameters of late-Victorian sexual culture.

The condition of 'living in' for shop assistants and most bar-maids, throwing together women from a range of class positions with varying degrees of sexual experience, is also presented by Gissing as having a negative effect on female respectability. As a practice it was seen by social investigators as unsatisfactory. An 1894 report commented on shop assistants' discontent with their 'bad domestic and sanitary arrangements' (Bulley, p.44), and a later report revealed that 'their moral calibre is weakened by the absence of proper social or home life' (Bondfield, p.277). Eliza Orme found that although the evidence given by bar-maids about their poor sleeping accommodation was 'untrustworthy', they were indeed often 'crowded in their bedrooms' with other servants (Bulley, pp.46f.), which may have fostered indecent conversations. In *The Odd Women*, Gissing seems more interested in the shop-girl's experience of living in than her potential exploitation within the shop (Sanders, pp.200f.). The shop-girl attracted sympathy because of the 'danger and fruitful source of deterioration in this career', as her 'desire for amusement' made her prey to all the 'attractions and temptations' of the town (Bird, p.69). Monica's employers encourage shop-girls to walk the streets in the evenings as well as on Sundays by providing them with latchkeys, a policy shown to encourage casual sex and prostitution. According to Clementina Black, the prospect of dismissal could also see typically 'shut-in' workers turned out into the streets at a moment's notice by uncaring employers (Black, p.54). The 'lively conversation' of the five shop-girls sharing a dormitory with Monica who discuss their sexual affairs with perfect freedom is 'little to her taste' (pp.48; 25), but like them, she then becomes accustomed to being approached in the street by men, adding to the novel's construction of shop-girls as 'sexually and morally suspect denizens of London' (Ledger, p.166). Cross-examined by Rhoda about her work, Monica claims that only one or two of her fellow workers are 'nice quiet girls' (p.35), so that her adoption of the freedoms of the much more typical 'showily-dressed' Miss Eade, later to become a prostitute, is easier to understand. Gissing's comment on the loudest of the shop-girls who returns at one in the morning, with her 'scandalous' anecdotes and 'innocent vulgarity', is particularly revealing: despite her 'morally unenviable reputation' (p.48), she is still envied by some of her colleagues. Rather than representing the encounters with men which would have taken place within the shop, Gissing chooses to highlight the dangers of the living-in system, which puts the shop-girl's morality in jeopardy through corruption from her peers and occupation of the streets.

It is in his examination of women's encounters with men in the city that Gissing is at his most ambiguous. Feminist historians have offered alternative readings of women's occupation of urban space. While Walkowitz stresses the respectable woman's vulnerability in the 'negative environment' of London (1992, p.46), Lynda Nead has argued for 'a repatterning of sexuality in the spaces of the city' and the 'bold new type of metropolitan femininity' this may have produced (Nead, pp.78, 131). Kathy Peiss's discussion of working women and leisure in turn of the century New York is based on a model of the 'mixed-sex fun' of heterosocial culture, which allows men and women to approach each other more easily (Peiss, pp.6ff.). Although Gissing did not explicitly endorse the 'repatterning of sexuality' associated with the new heterosocial spaces of London, his treatment of women's freedom of movement in the city appears generally sympathetic. Both the shop-girl and the musician are shown to value their urban mobility, with the shop-girl making the most of her compulsory day off in an 'extremely limited social life' (Bondfield, p.277) and the musician relishing her need to attend concerts and publicise her recitals in the fashionable West End.

Women's encounters with men raise questions about the contradictory class identities and levels of respectability associated with particular professions. Alma's position as a 'lady performer' nevertheless put her on a level with other women of dubious sexual and social status: Dymes tries to persuade her to come out as a singer, a figure seen as closer to the actress and like his other protégée, Ada Wellington, perhaps 'a trifle loud' (*Whirlpool*, p.257). The difficulties of classifying women workers were particularly pertinent for shop-girls; Bird claimed that 'every kind of girl, drawn from every class and strata of society, is found behind the counters of the modern emporium' (Bird, p.65). Although in conventional class terms Monica is 'far from a representative shop-girl' (*The Odd Women*, p.238), her behaviour is nevertheless classified as sexually forward. When she accepts Widdowson's attentions in Battersea Park, it is said that ' "she's picked up some feller today" ' (p.48). Sally Ledger argues that the shop-girl in *The Odd Women* 'inhabits the metropolis in a more self-confident and disruptive way' than the feminist heroine Rhoda, and credits her with a modern awareness of the 'dangers facing an unaccompanied single woman who befriends a man unknown to her' (Ledger, pp.164f.). Some working women at the turn of the century were fascinated by prostitutes and envied their 'sexual expressiveness', as Peiss has noted, copying their style of dress and 'play[ing] with the subculture of prostitution' (Peiss, p.65). It is significant that Rhoda's attempts to persuade Monica of the merits of clerical work prompt an appreciation of her sexual freedoms as a shop-girl: she fears the office as 'a worse form of bondage' and begins to relish her mobility in London: 'she felt something of the recklessness which formerly excited her wonder when she remarked it in the other shop-girls' (*Odd Women*, p.38). The 'recklessness' of her flirtations with Bevis and Barfoot after her marriage is marked by her indiscreet occupation of urban space; the 'chance encounter' (p.222) in the streets, outside Bevis's flat and in the art gallery. 'Being seen' in public places (p.247) is always liable to be misinterpreted as evidence of an illicit planned liaison. Female sexual agency, like that of another notorious shop-girl Amy Drake who seduces Barfoot on a train, is then seen as a variation on the 'vulgar' behaviour of Miss Eade, whose aggressive attempts to compete with Monica for male attention are presented as the undesirable product of a system that encourages women to befriend unknown men. The city certainly allowed women to express their sexualities in new ways, though Gissing implies that unaccompanied young women could be playing a dangerous game.

Entering 'the maelstrom of London life' might be acknowledged to be necessary for the woman of the future: even a largely conservative discussion of working ladies in the *Fortnightly Review* of 1892 conceded that 'there can hardly be too much freedom in one way in a woman's life' (Phillipps, p.200). However, the freedom reluctantly granted to the working woman rarely appeared compatible with marriage. The urban lifestyles dissected in the press mainly reflected the experiences of single women at a time when studies of marriage patterns increasingly indicated that the spinster, particularly in the capital city, was a growing social phenomenon (Collet, p.540). Alma's choice of profession ensures a far from typical working experience – she has neither fellow workers nor workplace in a conventional sense – but, crucially, her interest in music has to be developed in attendance at concerts, which demands her regular occupation of the male-dominated public sphere. As she exchanges her role as glorified spinster for suburban housewife before the end of the first volume, her attempts to use the city to further her career – 'it seemed to Alma that the very best thing for her health would be to spend a week or two in London and see her old friends, and go to a few concerts' (*Whirlpool*, p.170) – are inevitably seen as compromising. Attending a concert at the Crystal Palace is far from innocent when Redgrave, with his 'seductive familiarity', makes her into a sexual object: 'his eyes declared ... the sensuous pleasure of sitting by Alma's side' (p.246). Comparing his heroine to the single Leach sisters, Gissing remarks ambivalently:

> She ... ran up to town at least as often as they [Dora and Gerda] did. Like them, she found it an annoyance to have to rush to the station before midnight; but, being married, she could allow herself more freedom of movement than was permissible to single young women and having once missed the last train, she simply went to a hotel where she was known, and quietly returned to Pinner next morning. That Mrs Rolfe had such complete liberty and leisure seemed to them no subject for remark.... She had the precious advantage of being able to use London for all legitimate purposes without danger of being drawn into the vortex of idle temptations. (pp.188ff.)

Travelling by train alone to concerts and recitals is seen to be inappropriate and dangerous, as well as liberating; like the prostitute Alma occupies the streets around midnight and is 'known' in hotels, as Gissing qualifies the 'freedom of movement' of the professional woman. Her belief in the 'completeness' of her liberty is quickly revealed to be rather naïve, given the fretting husband waiting at home. However, Gissing is perhaps less consistently ambivalent about 'women's new freedoms' than Deborah Parsons suggests. The texts do not always tend to connect 'the independent and mobile woman' and her use of public transport with 'loose sexual morality' (Parsons, p.51): the Leach sisters' acceptance of Alma's freedoms and their contempt for her unsupportive husband suggest a more modern outlook. In Gissing's fiction, women's exposure to danger in their supposedly 'legitimate' use of the city is treated more seriously once they are married.

The retreat from the 'public life'

In an article written in 1900, Mary Jeune linked 'The Decay of the Chaperon' to the 'new relations' of a heterosocial culture, welcoming the move from supervision to self-reliance when a woman could 'take care of herself' (Jeune, p.637). Restrictions on women's freedom of movement in the later chapters of Gissing's novels, however, appear to reinforce the view that they must be protected from the sexual dangers of the public sphere. Like the prostitute Carrie Fisher in *Workers in the Dawn*, both Monica and Clara are stalked and spied on by their husbands and admirers, even when they have given up work in order to marry, as if women who have worked in such professions cannot be trusted to take care of themselves. Clara's anger at Sidney for 'daring to act the spy' outside her new lodgings (*Nether World*, p.95) mirrors Monica's contempt for her husband when he hires a private detective to shadow her movements. Clara's later facial disfigurement ensures a restriction on the mobility Sidney fears, and allows her to be safely repositioned in the home. This is explicitly related to reducing her marketability and exposure to sexual danger: her father typically experiences 'the joy of having [her] under his roof, rescued alike from those vague dangers of the past and from the recent peril' (p.244). By the closing chapters, we are told she 'never went out to make purchases' (p.373), only leaving the house heavily veiled under Sidney's chaperonage.

Rolfe's decision to chaperone Alma to her recital suggests a similar refusal to accept her new-found freedoms, and her forced acceptance of domestic confinement so that she has 'less temptation to gad about' (p.346) illustrates the workings of the same paradigm. According to an article on 'Women's Work: its Value and Possibilities' published in the *Girl's Own Paper* in 1894, the modern working woman, a 'commonplace' at the end of the century, was seen to be 'more admirable than she, who ... spends her days in domesticity' ('FH', pp.51f.); but this generally progressive account, like Gissing's texts, is shot through with warnings about unwomanliness. Ledger makes the valid point that Widdowson's 'paranoiac desire to closet Monica away from the public spaces of the city derives at least in part from the fear that any woman who shows herself in public might turn out to be a "public woman" ' (Ledger, p.165). However, it should also be noted that male fears around women 'showing themselves' in public are more likely to be activated in Gissing's fiction by women whose professions have already implicated them in a process of display.

On the other hand, the conclusions to the novels can also be seen to highlight the difficulties of sustaining the very 'public lives' (Davis, p.69) women workers in such professions were obliged to lead. This has often been read in terms of Gissing's conservatism, as critics have emphasized his apparent endorsement of anti-feminist medical views on women's unfitness for the public sphere. William Greenslade argues that 'Alma's neurosis is the inevitable "tax" she must pay for spending her energies recklessly in the public sphere' (Greenslade, p.149). However, little account has been taken of the economic factors which drive the performing woman back into the home: typically Gissing's borrowings from the social investigators suggest a more complex, ambiguous representation of the working woman's struggles. The experienced musician in Jeune's 1893 study of employment for ladies weighed 'the attractions of a public life' against 'the struggle to get engagements' (Jeune, p.41). As employment manuals repeatedly warned, 'the profession of a public performer is very precarious' (Philipps, p.97), defined by fierce competition, limited demand and slim chances of fame and success. Most musicians were obliged to teach to supplement their earnings in 'a career so hazardous' (Bird, p.217). Actresses with only average talent were advised not to enter a profession already 'overcrowded with mediocrities' (Philipps, p.2). Alma's attempts to progress from amateur to professional status as a musician appear rather misguided in the context of such warnings: her fear that she might 'c[o]me out and ma[k]e a fiasco' (*Whirlpool*, p.189) is only too perceptive. Similarly, the jealous throwing of vitriol by her fellow actress which cuts short Clara's promising career when she is 'so near success' (*Nether World*, p.286) can be seen in the context of hard-headed advice about the 'short-lived popularity' (Bird, p.241) achieved by 'beauty alone' in acting, underlining the very real difficulties of achieving success.

Jeune's claims for the legitimacy of acting are not borne out: it makes Clara 'less womanly', more open to 'degradations', until 'noble feeling was extinct in the girl' (*Nether World*, p.203). There is an attack on stereotypes about the sexual immorality of an actress's lifestyle in one of Clara's later speeches, in which she claims: ' "I know what people think about actresses ... do you imagine I have been living at my ease, congratulating myself that I had escaped from all hardships? ... I was working hard and honestly" ' (p.286); but she is permitted to voice such sentiments only after her withdrawal from acting. It is an ironic twist that her vulnerability on the streets is to a fellow actress rather than a predatory man. Grace's jealous actions follow her accusation that Clara has slept her way to the new part: ' "We know how creatures of your kind get what they want" ' (p.207). This may be read either as a comment on the competitiveness within the profession or as an extreme version of the aggressive sexual rivalry between the shop-girls in *The Odd Women*. The harshness of Clara's punishment is coupled with a sense of barely concealed satisfaction on the author's part that 'her career as an actress had ended' (p.243), giving her no option but to return to her father's home. Her acting skills can then only be used in the private realm to prompt Sidney's proposal of marriage rather than on the public stage where 'like all unadvised purchasers she had paid a price that might well have been spared' (p.281). In a chapter titled 'Woman and Actress', Gissing seems to suggest that the two are mutually exclusive and takes pleasure in restoring her womanliness by relocating her in suburban domesticity.

Bird warned that both the musician and the actress had to be strong 'to stand the strain of ... the late hours and excitement inseparable from public appearances' (Bird, pp.220, 241), acknowledging the 'exhausting' public life of the performer without trumpeting her unfitness for it. The ambiguity of Gissing's judgement of his violinist is compounded by an uncertainty about whether playing the violin for an 'at home' is a dangerous act: Alma's decision not to play to private audiences 'lest she should make herself cheap' (*Whirlpool*, p.255) recalls the scene where Rolfe first sees her, 'warm' and 'flushed' from her violin-playing until 'she stood a fascination, an allurement, to the masculine sense' (p.32). Alma's sense that her husband 'doesn't quite like the thought of my playing for money' (p.236) suggests an uneasiness about the public life of the professional which Gissing increasingly seems to share. Strangely, her first recital is not directly described, but the publicity leading up to it and its dangerous overtones are enough to ensure that it will not be repeated. Alma's 'glorious sense of power' in her 'musical triumph' (pp.313f.), rather than signalling her representation as a female vampire intent on seduction, as Gillett has claimed (Gillett, p.134), is more closely anchored in her deluded belief that freedom from her debt to Redgrave will guarantee that 'all danger was at an end' (p.310). The difficulties of achieving success signalled in her 'anxiety', 'agitation' and loss of energy have less to do with her suspect abilities and jealousy, as Kramer argues (Kramer, p.325), than with the sensational revelations in the newspapers about Redgrave's unexplained death and her fear of being named as the dead man's mistress. The risk to her health and the 'strain' of performance are inextricably linked to the sexual dangers to which she is exposed as a performing woman, ensuring that her future appearances 'at home' must necessarily be without her instrument.

Gissing's account of the sexual dangers of women's public lives can then be seen to borrow from the research of the social investigators and to respond to some of their warnings and concerns. Prevalent fears about sexual objectification and loss of respectability are shown to contribute to the short-lived nature of the performing woman's career, as she has to illustrate the 'deteriorating effects' of professions dependent on appearance. But the suburban marriage offered to the 'shut-in' women obliged to withdraw from the public sphere is a poor consolation. Both Alma and Clara have a special box of mementoes of their professional lives which they keep shut away, indicating their bitterness at having to relinquish their professional ambitions and dreams of success. Despite the increased opportunities of heterosocial culture, the novels ultimately show that professional women would be virtually unable to further their careers without the economic support of men who expect sexual favours in return. Bailey sets the risks of appearing in public against arguments about the bar-maid's palpable, if limited, control over her own respectability (Bailey, pp.160, 164). But Gissing's compromised women ultimately remain 'women at risk'.

Bibliography

Anon. (1896), 'Bar-maids and Waitresses in Restaurants, their Work and Temptations', *Girl's Own Paper*, 22 February 1896, pp.329-330.

Bailey, Peter (1990), 'Parasexuality and Glamour: the Victorian Bar-maid as Cultural Prototype', *Gender and History*, 2:2, pp.148-172.

Bateson, Margaret (1895), *Professional Women upon their Professions: Conversations Recorded by Margaret Bateson*, Horace Cox, London.

Bird, M. Mostyn (1911), *Woman at Work: A Study of the Different Ways of Earning a Living Open to Women*, Chapman and Hall, London.

Black, Clementina (1907), *Sweated Industry and the Minimum Wage*, Duckworth and Co., London.

Bondfield, Margaret (1899), 'Conditions under which Shop Assistants work', *Economic Journal* 9, pp.277-286.

Bulley, A. Amy (1894), 'The Employment of Women: The Lady Assistant Commissioners Report', *Fortnightly Review*, 61, pp.39-48.

Chialant, Maria Teresa (2001), 'The Feminization of the City in Gissing's Fiction: The Streetwalker, the *Flâneuse*, the Shopgirl' in Bouwe Postmus (ed.), *A Garland for Gissing*, Rodopi, Amsterdam.

Collet, Clara (1892), 'Prospects of Marriage for Women', *Nineteenth Century* 31, pp.537-552.

Davis, Tracy C. (1991), *Actresses as Working Women: their Social Identity in Victorian Culture*, Routledge, London.

Dreiser, Theodore, ed. Alfred Kazin (1981; first publ. 1900), *Sister Carrie*, Penguin, Harmondsworth.

'F.H.' (1894), 'Women's Work: Its Values and Possibilities', *Girl's Own Paper*, 27 October, pp.51-53.

Gillett, Paula (2000), *Musical Women in England, 1870-1914: 'Encroaching on Men's Privileges'*, Manchester University Press, Manchester.

Gissing, George, ed. Stephen Gill (1992; first publ. 1889), *The Nether World*, Oxford University Press, Oxford.

Gissing, George, ed. Margaret Walters (1980; first publ. 1893), *The Odd Women*, Virago, London.

Gissing, George, ed. Gillian Tindall (1984; first publ. 1897), *The Whirlpool*, Hogarth, London.

Greenslade, William (1994), *Degeneration, Culture and the Novel, 1880-1940*, Cambridge University Press, Cambridge.

Grein, J.T. (n.d.), 'A Bar-maid', in *Fin de Siècle: A Selection of late-nineteenth-century literature and art* (1947), ed. N.Wallis, Allan Wingate, London.

Grylls, David (1986), *The Paradox of Gissing*, Allen and Unwin, London.

Hardy, Thomas (1978; first publ. 1895), *Jude the Obscure*, Penguin, Harmondsworth.

James, Simon J. (1997), 'Negotiating the Whirlpool', *Gissing Journal*, 33, pp.15-25.

Jeune, Mary (1900), 'The Decay of the Chaperon', *Fortnightly Review*, 74, pp.629-638.

Jeune, Lady (ed.) (1893), *Ladies at Work: Papers on Paid Employment for Ladies by Experts in Several Branches*, A.D. Innes & Co, London [reprinted, after revision, from *The Monthly Packet*].

Kramer, David (2000), 'George Gissing and Women's Work: Contextualizing the Female Professional', *English Literature in Transition*, 43:3, pp.316-330.

Ledger, Sally (1997), *The New Woman: fiction and feminism at the fin de siècle*, Manchester University Press, Manchester.

Levine, Philippa (1987), *Victorian Feminism, 1850-1900*, Hutchinson, London.

Mattheisen, Paul F., Young, Arthur C. Young and Coustillas, Pierre (eds.) (1994-5), *The Collected Letters of George Gissing: Volumes 5 and 6*, Ohio University Press, Athens (Ohio).

Morley, Edith J. (ed.) (1914), *Women Workers in Seven Professions: A Survey of their Economic Conditions and Prospects*, edited for the Studies Committee of the Fabian Women's Group, Routledge, London.

Nead, Lynda (2000), *Victorian Babylon: People, Streets and Images in Nineteenth-Century London*, Yale University Press, New Haven & London.

Nord, Deborah Epstein (1995), *Walking the Victorian Streets: Women, Representation and the City*, Cornell University Press, Ithaca and London.

Parsons, Deborah L. (2000), *Streetwalking the Metropolis: Women, the City and Modernity*, Oxford University Press, Oxford.

Peiss, Kathy (1986), *Cheap Amusements: Working Women and Leisure in Turn-of-the-Century New York*, Temple University Press, Philadelphia.

Philipps, Mrs (1898), *A Dictionary of Employments open to Women*, Women's Institute, London.

Phillipps, Evelyn March (1892), 'The Working Lady in London', *Fortnightly Review*, 58, pp.193-203.

Robinson, F. Mabel (1887), 'Our Working Women and their Earnings', *Fortnightly Review*, 48, pp.50-63.

Sanders, Lise Shapiro (2001), 'The Failures of the Romance: Boredom, Class, and Desire in George Gissing's *The Odd Women* and W. Somerset Maugham's *Of Human Bondage*, *Modern Fiction Studies*, 47:1, pp.190-228.

Walkowitz, Judith R (1992), *City of Dreadful Delight: Narratives of Sexual Danger in Late-Victorian London*, Virago, London.

Walkowitz, Judith R. (1998), 'Going Public: Shopping, Street Harassment, and Streetwalking in Late Victorian London', *Representations*, 62, pp.1-30.

Chapter 3

'Just a morsel to stay your appetite': Gissing and the Cultural Politics of Food

Scott McCracken

> She passed on and entered the place of refreshment that was kept by Mrs. Tubbs. Till recently it had been an ordinary eating-house or coffee-shop; but having succeeded in obtaining a license to sell strong liquors, Mrs. Tubbs had converted the establishment into one of a more pretentious kind. She called it 'Imperial Restaurant and Luncheon Bar'. The front shone with vermilion paint; the interior was aflare with many gas-jets; in the window was disposed a tempting exhibition of 'snacks' of fish, cold roast fowls, ham-sandwiches, and the like; whilst further back stood a cooking-stove, whereon frizzled and vapoured a savoury mess of sausages and onions. (*The Nether World*, pp.30f.)

Food is at once the most basic necessity and the most mediated of cultural forms. It can be used in fiction to represent life at its most simple – daily bread, a bowl of rice, the humble potato – or at its most sophisticated, as haute cuisine. It can attract to the point where it becomes the object of our greatest desires, or repel and disgust. What is a delicacy in one culture can be taboo in another. Within cultures, distinction can be established by the status of different kinds of food, sliced white bread versus a wholemeal loaf, for example. Political struggles can be fought over or through food. Boycotts of Californian or South African grapes, or the choice of fairtrade coffee each translate received tastes into political actions that are all the more salient because they connect to visceral denials or satisfactions. Where we eat and how we eat are as important as what we eat. The trashing of a McDonalds or a Starbucks is a significant act because restaurants and cafés are points at which the public and the private (including intimate relations – the romantic meal) are negotiated. What is at stake in the destruction of such symbolic spaces is not just the Americanization of culture, but a sense of a civility lost to the processes of modernization. Italy's 'slow food' movement, to which Gissing might have been an ardent adherent, offers an alternative model of modernity, where taste, the distinctiveness of local or regional production and a different vision of the public sphere combine to re-establish family or community relationships in the face of the pressures of globalisation (see Slowfood).

The place of food in the culture of late nineteenth-century Britain was a long way from today's obsession with public, journalistic, and televisual preparation and consumption (Probyn, 2000). But some of the tendencies that make food an integral part of commodity culture were not only apparent, but well underway. While London had been the belly of the Empire for three hundred years, the building of railways into the interior of colonial territories (Asia and Africa, Canada, Australia), former colonies (the United States) and client states (South America) accelerated the supply and increased the quantities of foodstuffs available to the imperial metropolis.

In *The Nether World*, the (aptly named) Mrs Tubbs signifies that abundance as figure of plenty or female John Bull, while her renamed establishment indicates a patriotic pride in its origin. As important as its excess is its display. In the thematics of the novel the shopfront exhibits national prosperity. In its small way, it is a kind of Crystal Palace of food. This shift from production to consumption was indicative of a change in the balance of power in the food economy. The abundance of supply of tobacco, tea and grain meant that retailers gained the upper hand, playing off producers against one another, then selling at cut prices through outlets that attracted consumers, now accustomed to an exhibition culture, through ostentatious shop windows (Bird, pp.2-11).

The late nineteenth-century culture of abundance was also one of lack. In *The Nether World* nothing is given, nothing promised – whether the attractions of the stage, the hopes of marriage, or the Bank Holiday illusion of prosperity – that is not then taken away. At the Angel, Clara's entrance onto the stage of commodity spectacle marks Sidney Kirkwood's exclusion, just as the riches of the griddle mark his poverty. He reacts with repugnance to the 'savoury mess', a reaction which, according to David Trotter, is a response to 'the loss of self-respect entailed by what he regards as an immersion in matter' (Trotter, p.252). But food in Gissing's fiction represents more than the binary opposition of lack versus plenty and there is as much delight in the immersion in matter as disgust. In fact, food might be taken as a paradigmatic element of Gissing's realism: its necessity signifies the text's adherence to the real, but the basic need masks the performative aspect of an apparently determinist narrative.

In *Thyrza*, for example, the most vibrant representation of working-class life is the description of Lambeth market:

> On the outer edges of the pavement, in front of the busy shops, were rows of booths, stalls, and barrows, whereon meat, vegetables, fish, and household requirements of indescribable variety were exposed for sale. The vendors vied with one another in uproarious advertisement of their goods. In vociferation the butchers doubtless excelled; their 'Lovely, lovely, lovely! and their reiterated 'Buy, buy, buy!' rang clangorous above the hoarse roaring of costermongers and the din of those who clattered pots and pans. (*Thyrza*, p. 37)

Food, eating, and eating out operate within a rich matrix of meaning in Gissing's work and it would probably require a full-length study to explore all their aspects. David Trotter's section on food in Gissing's fiction in *Cooking with Mud* has begun the task with a fine account of the social distinctions Gissing draws between the uses of food in working-class culture (Trotter, pp.240-257). This chapter will attempt to map out the likely co-ordinates of such a study, using an unscientific selection of examples from *The Nether World*, *Thyrza*, *New Grub Street*, and *Born in Exile*, before concluding with a reading of *Eve's Ransom* as a work where the metaphors of food and eating structure the narrative. These co-ordinates include: 1) social Darwinism; 2) notions of bodily pleasure, Bacchanalia or *jouissance*; 3) manners, civility, the meal as socially symbolic act; 4) eating out and the boundary between public and private spheres; 5) the industrialization and commercialization of food and food culture. However, as will become clear from my examples, a proper analysis of Gissing's prose must lead us to conclude that it is not possible to separate these aspects artificially or to work through them systematically. Instead, it demonstrates, as has been ably argued by Simon James, that his novels can be read both for the determinations of plot and for a rich textuality that exhibits something closer to Bakhtin's dialogism than the scientific definitions one might expect from naturalism (James; Bakhtin, pp.410f.; Jameson, pp.202f.).

Darwinism and pleasure

An investigation into the culinary belies Gissing's reputation as a dour and serious writer. Focusing on food reveals one of the best kept secrets of Gissing's work – its humour. In an earlier passage from *The Nether World*, social Darwinism gives scientific authority to social prejudice, but that authority is subverted by Clem Peckover's robust enjoyment of both food and power. In what Trotter (Trotter, p.257) calls 'one of the finest moments not only in Gissing's writing, but in the totality of nineteenth-century representations of the nether world', a 'blow-out' of sausages brings a double opportunity to torture the hapless Jane Snowdon – both through violence and the withholding of food:

> "Ah!" exclaimed Miss Peckover (who was affectionately known to her intimates as Clem), as she watched Jane stagger back from the blow, and hide her face in silent endurance of pain. "That's just a morsel to stay your appetite, my lady! ... Ha ha! ho ho. These sausages is done; now you clean out that fryin'-pan; an if I can find a speck of dirt on it as big as 'arlf a farthing, I'll take you by the 'air of the 'ed and clean it with your face, *that's* what I'll do!" (*The Nether World*, p.6)

A blow to the head and her control of the food supply confirm Clem's superiority in the social order. We are told that:

> Civilization could bring no charge against this young woman; it and she had no common criterion. Who knows but this lust of hers for sanguinary domination was the natural enough issue of the brutalising serfdom of her predecessors in the family line of Peckovers? A thrall suddenly endowed with authority will assuredly make bitter work for the luckless creature in the next degree of thraldom. (p.6)

Clem exists in a state of nature, in which, according to nineteenth-century evolutionist science (and, earlier, Malthusian economics), access to nutrition ensures and perpetuates dominance. In *The Evolution of Sex*, published in the same year as *The Nether World*, Patrick Geddes and Arthur Thomson put digestion at the centre of the life-process:

> On the one hand, more or less simple dead matter or food passes into life by a series of assimilative ascending changes, with each of which it becomes molecularly more complex and unstable. On the other hand, the resulting protoplasm is continually breaking down into more and more simple compounds, and finally into waste products. The ascending, synthetic, constructive series of changes are termed 'anabolic,' and the descending, disruptive series, 'katabolic'. (Geddes and Thomson, pp.87f.)

According to Geddes and Thomson, even sexual difference can be understood in terms of metabolizing food. The processes of anabolism (building up) and catabolism (breaking down) mark the difference between female and male biologies:

> Femaleness is anabolic preponderance in reproduction, hence the ovum has necessarily the general character which this 'diathesis' produces in non-reproductive cells; and, similarly, katabolic preponderance stamps its character of active energy upon spermatazoa as naturally as upon the ciliated cell or the monad. (p.125).

The sperm is described as hungry, as looking for nutriment (125f.), and Geddes and Thomson conclude: 'Historically, then, fertilization is comparable to mutual digestion, and, though bound up with reproduction, has arisen from a nutritive want' (p.162). So unchallengeable was the social Darwinist discourse of nutritional necessity that feminist writers like Olive Schreiner and Charlotte Perkins Gilman developed their theories of women's subordination in relation to it. In *Woman and Labour*, Schreiner argued that bourgeois women's 'parasitism' was a consequence of their exclusion from productive labour (Schreiner, 1911). Gilman was more explicit about the role of nutrition in the 'sexuo-economic' relationship between men and women: 'We are the only animal species in which the female depends on the man for food, and the only animal species in which the sex-relation is also an economic relation' (Gilman, pp.4f.).

Feminist accounts of the relationship between nutrition and sex attempted to construct an ethical or political sphere beyond the natural where women could act to establish more equal gender relations. Gissing's own relationship with Darwinist science was similarly fraught (Ogden). This was partly because he also disliked the reductionist impulse in much social Darwinism, but also because he could not resist the polyvalent meanings in language that strove to be objective and truthful. Trotter argues that Clem 'eludes metaphor by the sheer *gusto* with which she attacks her food' (Trotter, p.256). Her description offers both a discourse of biological determinism and a delight in her natural bodily health and vitality: 'tall, rather bony, rudely handsome; the hand with which she struck was large and coarse-fibred, the muscles that impelled it vigorous' (*The Nether World*, p.5). Once explained as the product of evolution, Clem almost attains the status of noble savage: 'The frankness of Clem's brutality went far towards redeeming her character' (p.6). The reader is further discouraged from taking the passage as a serious social commentary by Clem Peckover's name, which, as so often in

Gissing's texts, acts as a self-conscious reflection on the character's theatrical role in the text, signifying performance rather than function. Clem is at the top of a pecking order, from where she preys on lesser beings like Jane and Pennyloaf.

Despite the enjoyment to be gained from such passages, it has been a commonplace of Gissing criticism that his work elicits few if any pleasures for the reader. Jane Miller, for example, says that comic moments are 'rare' (96): 'Gloom has always been the quality people have associated with Gissing's novels' (Miller, p.175). John Carey describes a moment of 'broad humour' as unusual (Carey, p.95). Fredric Jameson is a rare exception, commenting on the 'electrical dryness' of the later novels' 'wit' (Jameson, p.203). Yet even Gissing's bleakest ethnographic gaze offers a morbid enjoyment of the darkest 'lurking holes' of London's poorest districts. The (albeit vicious) humour visible in Clem's slapstick assault on Jane is just one end of a comic spectrum that extends from the sardonic mockery of the 'Io Saturnalia' chapter of *The Nether World* to the more subtle domestic dramas of the novels of the 1890s; and even a novel like *The Odd Women* contains some fairly crude sexual jokes. The virile hero is named Everard, the celibate heroine, Nunn, while the dried up, failed husband, Widdowson, retires to live with a friend called Newdick. More often than not Gissing's humour is dependent on a dialogic relationship between a concept of brute necessity and a comedy of manners, in which the rituals of food play a central role. Food is a signifier for economic necessity in Gissing's tales of impoverishment and near starvation, whether he is writing about working-class families (like the Hewetts in *The Nether World* or the inhabitants of Lambeth in *Thyrza*) surviving on a minimal diet, or struggling writers starving in their garrets. But while brute necessity haunts his texts as the reality of poverty in nineteenth-century London, he never neglects the symbolic role of food in the rituals of everyday life.

The meal as socially symbolic act

Maud Ellmann has argued that hunger, like sexuality, is a 'culturally constructed drive' (Ellmann, p.36). Just as the detailed account of the place of Bob Hewett's collar in the hierarchy of manual, semi-skilled and skilled workers in *The Nether World* anticipates twentieth-century cultural anthropology (Trotter, p.253), Gissing's work also anticipates that of Mary Douglas or Claude Lévi-Strauss in its understanding of the meal, or what anthropologists call the 'food event', as socially symbolic act (see Douglas). If food deprivation is represented as a sign of evolutionary failure, and eating well a sign of success, this is just the starting point for Gissing's investigation of the rituals of food and eating, where the 'meal is 'a microcosm of wider social structures and boundary definitions' (Lupton, p.9). In *New Grub Street*, for example, the history of Alfred Yule's marriage is presented initially in terms of an exact homology between access to nutrition and the possibility of reproduction:

> Living in a garret, and supplying himself with the materials of his scanty meals, he was in the habit of making purchases at a little chandler's shop, where he was waited

> upon by a girl of no beauty, but, as it seemed to him, of amiable disposition. One holiday he met this girl as she was walking with a younger sister in the streets; he made her nearer acquaintance, and before long she consented to be his wife and share his garret. His brothers, John and Edmund, cried out that he had made an unpardonable fool of himself in marrying so much beneath him; that he might as well have waited until his income improved. This was all very well, but they might just as well have bidden him reject plain food because a few years hence he would be able to purchase luxuries; he could not do without nourishment of some sort, and the time had come when he could not do without a wife. (*New Grub Street*, pp.123f.)

New 'grub' indeed (the slang term for food and the literary term come from the same root: to 'grub' is to dig either for hack work or for food, and of course for the former to get the latter). The closed economy of Yule's sexuality is neatly portrayed: a poor diet is determined by a low income, and thus limited purchasing power, in the growing culture of consumerism; consequently he has restricted access to a wider public sphere and his narrow social and sexual opportunities are represented through the metaphor of food – plain food versus luxuries. In such a reading, the passage invites a reductionist account of sexual desire as nothing more than brute instinct: a need that must be satisfied. But, while hunger and a Darwinian urge to reproduce function as the kernel of the real that authorizes a social commentary, the passage comes almost immediately after a more directly satirical account of Yule's appetites in the context of a family meal. Arriving home in a 'mood of silent moroseness', he finds that dinner is not on the table:

> Even the average man of a certain age is an alarming creature when dinner delays itself; the literary man in such a moment goes beyond all parallel. If there be added the fact that he has just returned from a very unsatisfactory interview with a publisher, wife and daughter may indeed regard the situation as appalling. (p.120)

The silent, family meal, subdued by the bad tempered patriarch, acts as a microcosm of the wider social structures of work, production (here literary production) and reproduction (the patriarchal family). Food and books perform a dialectic in Gissing's texts. According to Ellmann scarcity of food can be a spur to literary production (Ellmann, pp.18-21). In *New Grub Street*, Edwin Reardon 'keeps himself alive by writing, yet also immolates his body with his own prolixity, forfeiting his very substance to the public's greed for three-decker novels' (p.25). But books can also replace food as a source of enjoyment (p.22). In *Born in Exile*, Godwin Peak is 'compelled to contract his diet, that he might purchase books' (*Born in Exile*, p.52). In a passage in *The Private Papers of Henry Ryecroft*, the young Ryecroft forgoes food in order to buy the works of Tibullus:

> Many a time I have stood before a stall, or a bookseller's window torn by the conflict of intellectual desire and bodily need. At the very hour of dinner, when my stomach clamoured for food, I have been stopped by the sight of a volume so long coveted, and marked at so advantageous a price, that I could not let it go; yet to buy it meant pangs of famine. (*Ryecroft*, p.29)

During the Yule family meal, Alfred's daughter tries, unsuccessfully, to distract him from his hunger with a book. He only achieves satisfaction when, in the course of the meal, she tells him of the prospect of an editorship. Family harmony is restored to all when, buoyed up by the news, he comes upon a glowing reference to his work in the book he had previously rejected, but now finds sufficiently appetizing to read during the meal. Everyone gets their desserts when Yule remarks 'casually that the custard was very well made today' (*New Grub Street*, p.123).

The social Darwinist account of Yule's courtship is just one discourse placed in dialogic relationship with, first, Yule's own bodily consumption and, second, the central place of eating in the social symbolic. At first he eats hurriedly, without enjoyment: 'He began his meal by drinking half a glass of ale; then ate a few mouthfuls in a quick hungry way, his head bent closely over his plate' (p.120). When Marian mentions the editorship, Yule stops 'in the act of mastication. He fixed his eyes intently on the sirloin for half a minute; then, by way of the beer-jug and the salt cellar, turned them upon Marian's face' (p.121). On digesting the news, 'he continued to eat more slowly, and as if in appreciation of the viands'. Finally, after reading the good notice, 'he took a draught of ale, like one who is reinvigorated in the battle for life' (p.122).

Public consumption and commercial culture

Bodily need is in dialogic relationship with the social situation of the family and the wider public sphere, where Yule's literary reputation is made. In the home, Yule's brittle masculine authority is supported by his wife's culinary and his daughter's intellectual efforts. Outside the domestic sphere, the place of the male intellectual is far less secure. In contrast to Yule's 'plain food', Totty Nancarrow, in *Thyrza*, lives on luxuries:

> If she had not money enough for both bread and marmalade, she chose to have the marmalade alone; if she could not buy meat and pickles at the same time, she would have pickles and go without the meat. Marmalade and pickles she deemed the indispensables of life. (*Thyrza*, p.383)

Totty's taste for sweet and sour figures independent appetites that require a larger stage than the bourgeois home. In Gissing's early fiction that stage is the street, the market or the corner shop, urban spaces which give greater freedom to working-class women like Totty. In his later work, eating out in coffee shops and eating houses, and 'dining' in restaurants – at a time when opportunities for public consumption in all senses were widening – becomes a symbolic act in which individual human needs are mediated through larger social forces. In *Born in Exile*, a restaurant is the mark of shame that drives the hero, Godwin Peak, from public (academic) life. If the act of eating is an act of distinction in itself, his Uncle's planned establishment, 'Peak's Refreshment an' Dinin' Rooms' (*Born in Exile*, p.28), pronounced 'with the offensive purity of [a] Cockney accent' (p.24), represents the invasion of England's streets by the new commercialism.

Godwin's first reaction to his Uncle is bodily, a form of corporeal disgust: 'For an instant he was clay colour, then a hot flush broke upon his cheeks' (p.24).

His next response is to break away from his student friends and to try and confine the source of disruption to the private sphere, his lodgings; but Andrew Peak, acting like the force of commerce he is (a precursor of Luckworth Crewe in *In the Year of Jubilee*), impels them both into the semi-public, semi-private space of a teashop opposite the college. Godwin manages to direct his Uncle to a quieter, more private, room, but he cannot prevent his open questions to the waitress concerning the viability of the business. The dialogue between Uncle and nephew enacts, at a personal level, a battle for the high street. The teashop, like the chandler's shop where Yule encounters his wife, stands for an old and outdated artisanal economy: the waitress is asthmatic as if to emphasize the exhausted state of the business, the eating area almost empty. The period between 1874, when the novel opens, and 1900 saw the transformation of Britain's city streets. Individually owned shops and businesses began to be replaced with chains of retail outlets, teashops and restaurants. Andrew Peak remarks that what the area needs is a 'Spiers and Pond', a branch of one of the new successful London restaurant chains. The choice demonstrates Gissing's ever-accurate eye for historical and social detail. Spiers and Pond opened their famous Criterion restaurant in Piccadilly in the same year the novel opens, but Andrew Peak's admiration also indicates his social class. In W.S. Gilbert's Ballad, 'Fanny', the firm is associated with downmarket tastes:

> Fanny and Jenny in Paris did dwell,
> Miss Jane was a dowdy,
> Miss Fanny a swell –
> Each went to dine at a quarter to four –
> At her own little favourite Restauratore –
> Fanny of Bertram and Roberts was fond
> While Jenny she worshipped her Spiers and Pond

It is the prospect of his family name becoming a brand, of his Uncle's ambition to 'flare his name upon a placard' (p.60), that Godwin finds (physically) unpalatable: a piece of toast that he is 'endeavouring to masticate' as his Uncle unveils the idea 'turns to sawdust upon his palate'. Godwin's distaste stems from his resistance to being visibly associated with a burgeoning commercial culture, to becoming one of its signs. This is not just a question of being associated with Uncle's lower class origins, but of his sense of being incorporated into a system of exchange in which his intellectual successes will become one form of value amongst many. In his later novels Gissing came to see consumer culture as the medium through which all social relations are lived and this is nowhere more true than in his short novel, *Eve's Ransom*.

Eve's Ransom

The biblical story of the fall presents Eve with a prohibition that defies 'natural' law. The divine command not to eat takes food out of the sphere of nature and into that of culture. In *Eve's Ransom*, the basic necessity of eating is taken through a series of dialogic levels that correspond to the co-ordinates outlined above. If biological necessity stands as the first level (social Darwinist theories of nutrition), hunger in the story also stands as an implicit metaphor for sexual desire (bodily pleasure). The relationship between Hilliard and Eve is conducted through a series of 'food events', three of which involve meals, of which two involve eating out. These events define the boundaries of their relationship (the meal as socially symbolic act); but the semi-private, semi-public character of eating out situates the performance of a gendered identity at the threshold between private and public spheres. Finally, eating out itself is conducted in the midst of the new commercialized public culture of the two cities, London and Paris, in which the couple's encounters take place. Thus, Gissing's text explores the transformations of biological necessity through the limits of food's capacity to act as a form of display in consumer culture. The culture of consumption, as much as eating, becomes the narrative's theme.

Hilliard's first meeting is not with Eve herself, but with her image, a photograph belonging to his landlady in Birmingham, which he consumes in her absence. The moment has to be read in terms of the complex representation of the male gaze in Gissing's fiction. For example, in two of the novels of the 1890s, the heroines are presented to the reader posed on their beds in the tradition of the female nude. In *The Odd Women*, the image of Rhoda Nunn, writhing on the bed as she resists her desire for Everard, adds a sadistic element to that of voyeurism: 'she shed bitter tears; and not only wept, but agonized in mute frenzy, the passions of her flesh torturing her until she thought of death as a refuge' (*The Odd Women*, p.282). It is difficult to see this image as anything other than a form of revenge on the woman who dares to lead an independent public career. In *In the Year of Jubilee*, the representation of Nancy Lord, naked on a hot summer night, is explicitly sensual, but, as Arlene Young has pointed out, Nancy's sexual self-confidence is directly related in the passage to her new found public confidence at the Jubilee celebrations (*In the Year of Jubilee*, p.113; Young). Nancy is not just presented for the consumption of the desiring male reader. She is herself a consumer. The famous crowd scene in the novel offers the male reader an experience of cross-dressing, where he experiences the crowd through the eyes of a woman in a series of public encounters where her self-assurance does not falter. Nancy offers an early example of the female masculinity explored by Dorothy Richardson in Miriam, the heroine of *Pilgrimage* (McCracken).

The relationship between Hilliard and Eve does not offer so self-assured a heroine, but Eve combines Rhoda's resistance to masculine advances with some of Nancy's confidence in her own sexual power. The mystery that Eve represents is essentially the mystery of how much agency she possesses in the new urban culture, caught as she is between her position as (limited) consumer and as object of a consuming male gaze: in Rachel Bowlby's words, 'seducer and seduced, possessor and possessed' at the same time (Bowlby, p.32). The power difference between Hilliard and Eve is signified, as with Yule's social status, by their different purchasing power, endowed by the greater and lesser sums of money each unexpectedly receives. Eve and Hilliard's different positions within the new consumer economy are then played out at meals eaten out. As with the encounter between Godwin and Andrew Peak, the restaurant is the point at which public and private spheres intersect.

For Hilliard, the money he has received, four hundred and thirty-six pounds, buys him out of an alienated life, life as a machine: 'Before long there'll be machines for washing and dressing people – machines for feeding them' (*Eve's Ransom*, p.13). After yielding to 'London's grossest lures', his more virtuous life in Paris is figured by meals that offer good value. He eats 'at restaurants where dinner of several courses costs two francs and a half'. His mind turns to Eve's address, given to him by his landlady on leaving Birmingham, over a 'bowl of coffee and *petit pain*' (p.20).

Hilliard's first encounter with Eve is at the Health Exhibition at Kensington. The Victorian exhibitions were not only opportunities for commodity display, but for huge catering contracts competed for by firms like Spiers and Pond (which specialized in railway catering as well as restaurants and hotels) and later companies like J. Lyons and Co. These companies saw the new workforce of women as a new market looking for respectable dining at a reasonable price. Eve's autonomy is figured by sitting down at one of the open air tables, where (like Bloom in Davey Byrne's) she orders and consumes a glass of wine and a sandwich. For Hilliard, the pre-lapsarian Eve, 'the meek, the melancholy, the long-suffering, the pious', all but disappears in this new Eve who meets a man at the exhibition and willingly consumes the fruit that London has to offer: 'She had the look, the tones, of one bent on enjoying oneself, of one who habitually pursued pleasure, and that in its most urban forms' (p.25). If the tale is read as an updated account of the Fall, then this man, who has the 'external attributes of a gentleman', plays the part of the serpent, luring Eve away from the path of righteousness. However, this would be the scene as understood by Hilliard. We are told the man lacks the devil's attractions or, apparently, his ill intent: 'He was plain of feature, but wore a pleasant, honest look and his demeanour to the girl showed not only good breeding but unmistakable interest of the warmest kind' (p.27). This apparent benevolence throws Hilliard's position as moral judge into doubt. In fact, the passage suggests that they begin from a position of equality in the public sphere, where equality is figured by independent acts of consumption. Hilliard, who has followed her from her lodgings, orders a bottle of beer, which he drinks at a neighbouring table before plucking up the courage to approach her. At this point, Hilliard is, if anything, the lesser party, far less confident in the spectacular space of the exhibition, whereas Eve follows in the footsteps of Clem Peckover, Totty Nancarrow or Alice Mutimer in *Demos* in her new-found confidence in the city.

Their second meeting is at the theatre, another place to see and be seen. At their third, an encounter in the street engineered by Hilliard after more liquid courage, he invites her to dine. The restaurant combines the scale of exhibition catering – 'a great hall, where some scores of people were already dining' – with the possibility of privacy, 'a nook' (p.42). Whereas each consumed separately, but on an equal basis, at the exhibition, together the difference in their economic positions is enacted through their relationship to food. This time Eve's autonomy, in contrast to that of Hilliard, is expressed not through consumption but through abstinence: 'Hilliard began his dinner with appetite and gusto; the girl, after a few sips, neglected her soup and glanced about the neighbouring tables'; 'Hilliard drank a glass of wine and saw that Eve just touched hers with her lips' (pp.43f.).

Eve's reluctance to eat is at once a performance in tune with social Darwinist theories and, at the same time, a symbolic resistance to Hilliard's power. She is the passive, 'anabolic' body: Hilliard is the active, hungry, 'catabolic', male-in-pursuit. She is in a state of dependence on Hilliard for her nutrition that equates both with Gilman's 'sexuo-economic' relation and Schreiner's 'sex-paratisism'. But while the dining scene reflects Darwinist discourse it also subverts it. Eve's distracted glances to other tables indicate her dissatisfaction and the other possibilities the city offers. Hilliard's pompous celebration of dining as a form of higher civilization is gently punctured by Eve's humility. His grandiose plans of excess are countered by her practicality:

> When the fruit had been set before them – and as he was unpeeling a banana:
>
> "What a vast difference," said Hilliard, "between the life of people who dine, and of those who don't! It isn't the mere pleasure of eating and the quality of the food – though that must have a great influence on mind and character. But to sit for an hour or two each evening in quiet, orderly enjoyment, with graceful things about one, talking of whatever is pleasant – how it civilises! Until three months ago I never dined in my life, and I know well what a change it has made in me." (p.45)

In contrast to Hilliard's phallic display, 'unpeeling a banana', Eve deflatingly remarks on her continuing lack of access to what in London particularly was still a very masculine sphere of public life and civility.

> "I never dined till this evening," said Eve.
> "Never? This is the first time you have been in a restaurant?"
> "For dinner – yes." (p.45)

Eve's answer suggests that while she has taken advantage of the daytime opportunities now available to women, she has eschewed London's more dangerous nightlife. Where at least one influential writer on public culture sees silence in public as a retreat to the private sphere (Sennett, p.215), Eve's reticent response to Hilliard's advances suggests that abstinence and silence may also be a reaction to public inequalities that debar free and frank speech. Hilliard receives her brief statement with joy. If she has not dined out 'there could not have been much intimacy between her and the man she met at the exhibition'.

There is another, more political, reading of Eve's fasting. Totty Nancarrow boycotts her local shop in *Thyrza* as a protest against the owner's treatment of a tenant. A few years later, the hunger strike would become a part of the armoury of the women suffragists (Ellmann). Eve's resistance can be read as a reaction to the new economy that offered women the opportunity to consume, but no political rights other than a decision to abstain.

Their second meal takes place under different circumstances. Eve is now kept by Hilliard in Paris, the city that was, to paraphrase Benjamin, the capital of nineteenth-century consumerism. Hilliard tells of his recent trip to the Alps, 'whilst he nibbled an olive' (p.72). Eve's consumption or lack of it is not described, but she is far from reticent. Instead, she is hungry for news of the journey and in particular for amusing stories of Hilliard's rich friend (and her future husband) Narramore. This hunger replaces the distraction at the earlier meal and it becomes clear Eve has learnt, or gained sufficient confidence, to manage Hilliard's dominant position at the table:

> The dinner went merrily on, and when the coffee was set before them:
>
> "Why not have it outside?" said Eve. "You would like to smoke, I know." (p.73)

But when Hilliard takes this opportunity to tell Eve his prospects as an architect, she takes advantage of the new position she has engineered to avoid a direct response: she is 'gazing across the street' (p.73). Feminine autonomy is achieved through evasion and denial. A couple of pages later we are told that when she arrived in London, ' "I was obliged to dress decently, and that came out of my food" ' (p.75). When at the end of the this encounter 'her eyes drew him with the temptation he had ever yet resisted', her acquiescence is framed in terms of a negative: 'Eve did not refuse his lips' (p.80).

These limited opportunities for freedom recede when they both return to Dudley with its far less developed public culture. Eve lacks the employment opportunities open to her in London, and Hilliard feels far more secure about his relationship to her. While his hunger continues, hers recedes, and is replaced with a sense of obligation that enrages Hilliard:

> "Your gratitude be hanged! Pay me back with your lips – so – and so! Can't you understand that when my lips touch yours, I have a delight that would be well purchased with years of semi-starvation? What is it to me how I won you? You are mine for good and all – that's enough." (p.90).

Here, and at a key encounter over a meal, this time in Hilliard's lodgings where, significantly, she has no view of a sphere beyond the private, she fails to perform as in Paris: 'she pretended to eat, but in reality scarce touched her food' (p.94). She also refuses a gift of port given to Hilliard by Narramore. Again her desire for Hilliard mirrors her desire for food and once again it is the mouth and particularly the lips that figure her refusal: 'If she granted him her lips they had no fervour respondent to his own: she made a sport of it, forgot it as soon as possible' (p.96). Her self-control becomes a method for controlling Hilliard: 'Upon Hilliard's vehement nature this acted provocatively; at times he was all but frenzied with the violence of his sensual impulses. Yet Eve's control of him grew more self-assured the less she granted of herself; a look, a motion of her lips, and he drew apart, quivering but subdued' (p.96). Eve, in effect, spits out her lover before leaving him.

Eve's subsequent approach to Narramore is, in a Darwinian sense, a rational move. She goes at first for work (that is, to secure an independent income), when Hilliard refuses to act as an intermediary, but Narramore's attraction is clear at once. For Eve, the brass bedstead business, an almost absurd parody of bourgeois marriage (bolstered by Narramore's own ironic remark ' "if brass bedsteads keep firm" ': p.92), offers a better source of financial security than Hilliard's still uncertain prospects and volatile character. Narramore offers the narrow *amour* of middle-class matrimony. Eve is quite open about her choices: 'If I had been free I would have married him – just because I am sick of the life I lead, and long for the kind of life he offered me.... It's your mistake to think me a crafty plotting, selfish woman. I'm only a very miserable one...' (p.117). Released by Hilliard and married to Narramore she becomes almost unrecognizable: 'He... only just persuaded himself of her identity'. It is not clear whether this is because she is better fed or even, to fulfil the biological destiny designated by Geddes and Thomson, pregnant. Instead, the final chapter stresses the excellence of her performance: 'Mrs Narramore was perfect in society's drill' (p.123).

Eve's Ransom is not really a novel in length nor in terms of its narrative structure. It is rather a literary novella that takes a parable as its source. Eve's final appearance in an English garden is deeply ironic: despite her fall she has returned to, not left, Eden. The final image of the novel is of fruit, still unplucked: 'Thick-clustered berries arrayed the hawthorns, the briar was rich in scarlet fruit' (124). But the biblical parallels are only one intertext in a work that demonstrates the richness of Gissing's best fiction. The metaphor of eating structures the whole text and its meanings modulate between Darwinian discourse, the meal as social ritual and the new possibilities of the city streets, transformed by shops, restaurants and exhibitions that offer new opportunities for public interaction between the sexes. If the new commercialism is the latter's most inescapable determinant, then Gissing's dialogic imagination invites multiple readings. But this is not to say that he ever abandons a concept of the real, which persists in his insistence on hunger (and indeed desire) as basic necessity. Scarcity and the widespread adulteration of food are the other side of abundance for some. But a full exploration of the politics of food in his texts, which this chapter has only really begun, demonstrates his keen interest in a performative aspect to social life that extends from a response to basic necessities – plain food – to the, always brief, enjoyment of luxuries.

Bibliography

Bakhtin, M.M. (1981), 'Discourse in the Novel', in *The Dialogic Imagination*, University of Texas Press, Austin.

Bird, Peter (2000), *The First Food Empire: A History of J. Lyons and Co.*, Phillimore, Chichester.

Bowlby, Rachel (1985), *Just Looking: Consumer Culture in Dreiser, Gissing and Zola*, Methuen, London.

Carey, John (1992), *The Intellectuals and the Masses: Pride and Prejudice among the Literary Intelligentsia 1880-1939*, Faber, London.

Douglas, Mary (1998), 'Coded Messages', pp.103-110 in S. Griffiths and J. Wallace (eds.), *Consuming Passions: Food in the Age of Anxiety*, Mandolin, Manchester.

Ellmann, Maud (1993), *The Hunger Artists: Starving, Writing and Imprisonment*, Virago, London.

Geddes, Patrick and J. Arthur Thomson (1889), *The Evolution of Sex*, W. Scott, London.

Gilman, Charlotte Perkins (1898), *Women and Economics*, T. Fisher Unwin, London.

Gissing, George (1910; first publ. 1892), *Born in Exile*, Nelson, London.
Gissing, George (1980; first publ. 1895), *Eve's Ransom*, Dover Books, New York.
Gissing, George (1987; first publ. 1894), *In The Year of Jubilee*, Hogarth, London.
Gissing, George (1974; first publ. 1889), *The Nether World*, Harvester, Brighton.
Gissing, George (1968; first publ. 1891), *New Grub Street*, Penguin, Harmondsworth.
Gissing, George (1980; first publ. 1893), *The Odd Women*, Virago, London.
Gissing, George (1987; first publ. 1903), *The Private Papers of Henry Ryecroft*, Oxford University Press, Oxford.
Gissing, George (1984; first publ. 1887), *Thyrza*, Harvester, London.
James, Simon J. (2001), 'Experiments in Realism: How to read a George Gissing Novel', in Bouwe Postmus (ed.), *A Garland for Gissing*, Rodopi, Amsterdam, pp.11-21.
Jameson, Fredric (1981), *The Political Unconscious*, Methuen, London.
Lupton, Deborah (1996), *Food, the Body and the Self*, Sage, London.
McCracken, Scott (2001), 'From Performance to Public Sphere: the production of modernist masculinities', *Textual Practice*, 15, 1, pp.57-59.
Miller, Jane (1990), *Seductions: Studies in Reading and Culture*, Virago, London.
Ogden, Stephen (2001), 'Darwinian Scepticism in George Gissing's *Born in Exile*', in Bouwe Postmus (ed.), *A Garland for Gissing*, Rodopi, Amsterdam, pp.171-178.
Probyn, Elspeth (2000), *Carnal Appetites: Food Sex Identities*, Routledge, London.
Schreiner, Olive (1911), *Woman and Labour*, T. Fisher Unwin, London.
Sennett, Richard (1977), *The Fall of Public Man*, Alfred A. Knopf, New York.
Slowfood (2003): see the website, www.slowfood.com, last consulted by the author in August 2003.
Trotter, David (2000), *Cooking with Mud: the idea of mess in nineteenth-century art and fiction*, Oxford, OUP.
Young, Arlene (2003), 'Mapping the Citizen: Characterisation and the City in *The Odd Women*, *In the Year of Jubilee*, and *Eve's Ransom*, paper given at the Second International Gissing Conference, 'Gissing and the City', London, July 2003.

Chapter 4

The Strange Case of Godwin Peak: Double Consciousness in *Born in Exile*

Jenny Bourne Taylor

Something weird happens to Godwin Peak when he is invited to dine with the Warricombe family after meeting his old Whitelaw College colleague in Exeter. At first, with typical self-consciousness, he scrutinizes his speech and movements – 'Any defect of pronunciation, any native awkwardness of utterance?' – thankful that he 'defie[s] inspection' as far as his clothes and general performance are concerned. He is hopeful that he might break out of his narrow social existence as an analyst in a Rotherhithe chemical factory – that he might expunge the horrible memory of his uncle's 'Refreshment an' Dinin' Rooms', and could 'by the force of his personality, ... obliterate from their minds such disagreeable thoughts as they might secretly entertain' by his performance in the private theatre of middle-class domesticity (*Born in Exile*, pp.156f.). Suddenly he is asked about Sunday's sermon at the Cathedral, which he has half-heard while covertly contemplating Sidwell:

> Peak's reply was one of those remarkable efforts of mind – one might say, of character, – which are called forth, without premeditation, almost without consciousness, by a profound moral crisis. A minute or so ago he would have believed it impossible to recall and state in lucid terms the arguments to which, as he sat in the Cathedral, he had barely given ear.... [At] the time he was indisposed to listen seriously, and what chance was there that the chain of thought had fixed itself on his memory? Now, under the marvelling regard of his conscious self, he poured forth an admirable rendering of the canon's views, fuller than the original, more eloquent, more subtle.... Godwin felt that a slight bow of acknowledgement was perhaps called for, but not a muscle would obey his will. (pp.157f.)

Godwin Peak's powerful social ambition crystallizes into the desire to join the ranks of the traditional upper-middle class soon after this moment – a desire figured in his imagination as his evolutionary destiny; the means by which he could 'cast away the evil grudge, the fierce spirit of self assertion, and be what nature had proposed in endowing him with large brain, generous blood, delicate tissues' (pp.160f.). His resolution to impersonate a Broad Church Anglican in order to become intimate with the Warricombe family and marry a 'lady as England has perfected her at the close of this nineteenth century' (p.170) is made in response to this simultaneously physical and emotional drive, which has been performatively prefigured in the earlier moment of lapsed consciousness. Peak then rationalizes his decision to make real this spontaneous impulse, seeing it as the logical acting out of his relativist secular principles reinforced by a Darwinian model of sexual selection, and justifying it with his own brand of special pleading: 'Let the society which compelled to such an expedient bear the burden of its shame' (p.169). But his announcement to Buckland Warricombe that he is going to train for the clergy – again driven by forces apparently beyond his control – is followed by very different emotions; now he sees his unconscious self as the perverse acting out of a very different story:

> Since the display of rhetoric at luncheon all was downright somnambulism. What fatal power had subdued him? What extraordinary influence had guided his tongue, constrained his features? His conscious self had had no part in all this comedy; now, for the first time was he taking count of the character he had played.... Into what unspeakable baseness had he fallen? Happily, he had but to take leave of the Warricombe household and rush into some region where he was unknown. Years hence, he would relate the story to Earwaker....
>
> Yet such a possibility had not been unforeseen.... His imagination had played with the idea that a man like himself might well be driven to this expedient, and might even use it with life-long result....
>
> But he, he who had ever prided himself on his truth-fronting intellect, and had freely uttered his scorn of the credulous mob! He who was his own criterion of moral right and wrong! No wonder he felt like a whipped cur. It was the ancestral vice in his blood, brought about by over-tempting circumstance. The long line of base-born predecessors, the grovelling hinds and mechanics of his genealogy, were responsible for this. (pp.178f.)

It is only by the middle of the third part of *Born in Exile* that Godwin has started 'to review with calmness the course by which he had reached his now steadfast resolve' to deliberately cultivate this false persona, a resolve which is at once the expression of his 'vindicated will' and a recognition of internal fragmentation: 'Injury to himself he foresaw and accepted... he could never be the man nature designed in him; and he must frequently submit to a self-contempt which would be very hard to bear' (pp.215f.). And these splits in consciousness recur throughout the novel, as Godwin battles with the implications of his decision to 'live a life of deliberate baseness' in order to become a legitimate social subject (p.215). For all his attempts to reinvent himself, to 'knit himself into the social fabric', he is inexorably driven to further exile as he is brought face to face with his past (p.482). Moving between the wider group and an intense focus on Godwin's mental life, culminating in moments of crisis when he is brought face to face with traces of his past, this is Gissing's most complex investigation into the ways in which class relationships and intellectual institutions mould individual subjectivity and into the perverse, self-destructive psychic patterns they produce, as social divisions are taken apart via fissures within the mind.

In *Born in Exile* (1892), the fraught position of the 'unclassed' intellectual moves inward, into a study of the contradictory and perverse contortions of an individual consciousness, but a consciousness freighted with social significance. Like many of Gissing's ambiguous heroes, Godwin Peak is driven by what Martin Ryle describes in the present book as a 'spirit of antagonism'; but now it has become an openly paradoxical impulse of repulsion and attraction, as the hostility against his own lower middle-class family not only turns into a half-admiring, half-hostile envy of hereditary privilege, but is ultimately directed against himself. While Godwin's working-class contemporary John Earwaker manages to carve out a secure social niche and achieve an integrated position as a man of letters, uncompromised by either traditional hierarchy or mass culture, Peak himself personifies that conflict between intellectual radicalism and social conservatism which has made him one of Gissing's most intriguing and infuriating heroes. 'Part prig, part snob, all egoist, Godwin Peak inspires us with feelings where disgust and boredom strive for mastery; his talk, his thoughts are ever of himself', the *Saturday Review* commented in June 1892. While the *Daily Chronicle* was more sympathetic, quoting Matthew Arnold – Godwin 'stands "between two worlds, one dead, one waiting to be born" ' – it too noted that 'we don't like him of course, neither does his creator', though 'in spite of ourselves our dislike is tempered by more than a dash of genuine pity' (Coustillas and Partridge, pp.200ff.).

Twentieth-century readers have taken a broader view of the novel, placing Godwin's struggle to become 'his own criterion of moral right and wrong' within the cosmopolitan intellectual tradition of Bourget, of Dostoevsky, of Nietzsche (for example Korg; Swann). Yet as victim exposing the liberal myth of social mobility and social climber in collusion with the conservative hierarchy he attempts to scale, Peak highlights very English cultural conflicts (Alden; Goode; Sloan). Like Hardy's *Jude the Obscure* (1895), *Born in Exile* depicts aspirations deformed and frustrated by the residual weight of a still dominant culture; for all their differences of class and caste, both Jude and Godwin at first internalise the dominant values which exclude them – both see the Church as the means of social advance. But while Jude moves towards a more radical estrangement, the career of Godwin, as John Goode has argued, 'is simply a record of the failure to achieve integration'. He remains a morbid symptom of persistent social divisions in a narrative that is at once too detached and too identified with its central consciousness to produce any coherent critical distance from it (Goode, pp.60f.). Here the limitations of Godwin's consciousness are inextricably linked to *Born in Exile*'s failure to create a mode of self-consciousness able to finally transcend its own earlier delusions and hereditary determinations.

Yet it is precisely in exploring these 'morbid symptoms' that *Born in Exile* starts to anticipate modernist narrative: it dissects Godwin's self-absorbed and divided psyche, while placing him resolutely within a social world. David Glover discusses here how Gissing's novels of the 1890s anatomise kinds of nervous disorders (particularly of women) created by the pressures of late Victorian modernity similar to those Freud would investigate in his case histories. In *Born in Exile*, however, the critique of modern culture offered in *In the Year of Jubilee* or *The Whirlpool* is refracted through the personality of its central protagonist, in the most extended psychological case study to be found in Gissing's fiction. Rather than the heat and glare of fashionable society, the keenly observed social stratification of the London suburbs or the physiological depletion and psychic exhaustion of neurasthenic women, the story is focused on a lower middle-class man who is at once fragile and controlling, arrogant and insecure, and the group that immediately surrounds him. This is an intellectual generation, moreover, that is brought together and shaped by a particular institution – Whitelaw College – at a particular moment in the 1870s (this is arguably England's first campus novel).

Godwin is the product and victim of the tension between the enduring dominance of a residual caste of traditional intellectuals and an emergent scientific culture, a tension manifest in the social and intellectual divisions he encounters at Whitelaw College. But *Born in Exile* also shares the contemporary fascination (which Freud would draw on and transform) with the workings of the mind which was produced by that scientific culture itself: a fascination with how the self is shaped by memory, by unconscious impulse and conscious will.

These social and psychological divisions are brought together through the mid-nineteenth-century concept of *double consciousness.* Godwin first falls into a kind of trance on his encounter with the Moxey family immediately after returning to Twybridge after the prize-giving scene at Whitelaw College, under the pressure of trying to maintain two trains of thought (polite small talk, and speculation about Christian Moxey's taste in women). The narrator comments:

> Such play of the imaginative and speculative faculties accounts for the common awkwardness of intelligent young men in society that is strange to them. Only the cultivation of a double consciousness puts them finally at ease. Impossible to converse with suavity, and to heed the forms of ordinary good breeding, when the brain is absorbed in all manner of new problems: one must learn to act a part, to control the facial mechanism, to observe and anticipate, even whilst the intellect is spending its sincere energy on subjects unavowed. (p.71)

Godwin is briefly taken over by a 'second self' beyond his own control later in the novel, but in this passage his performative assumption of this divided identity is a direct expression of the internal splitting created by a conscious suppression which is the product of his lower-middle-class history and shored up by his experience at Whitelaw. Gissing thus gives 'double consciousness' itself a double meaning in *Born in Exile*, as these two forms of mental duality clash and correspond. He appropriates a well-known psychological term and reworks it in an explicitly social context as a form of class consciousness, while playing on its complex significance in psychological discourse to destabilize naturalized perceptions and assumptions. This dual meaning structures the novel's implied overarching epistemological framework and its narrative voice, as well as shaping the paradoxical consciousness of Godwin Peak himself.

'Cultivating a double consciousness': memory and social identity

Born in Exile shares with other fictional cases of the *fin de siècle* such as *The Strange Case of Dr Jekyll and Mr Hyde* and *The Picture of Dorian Gray* a fascination with what the co-founder of the Society for Psychical Research, F.W.H Myers, described in 1886 as 'the multiplex and mutable character of that which we know to be the Personality of man' (Myers, p.648). These comparisons are limited, and Gissing might well have resented them: of Stevenson, he wrote in his diary in 1887 'my prejudice against the man is insufferable, inexplicable, painful; I hate to see his name and I shall never bring myself to read one of his books' (though he later records reading his fiction: see Coustillas, 1978, p.33). However, when Godwin engages in two strands of thought simultaneously, or watches himself expressing unconsciously remembered ideas, he is participating, as patient and observer, in a form of mental dissociation that had fascinated mental scientists and been exploited by novelists such as Dickens and Collins as a realistic means of generating uncanny effects for the past forty years.

The belief that the self is formed through the interaction of conscious and unconscious memory had been rooted in the Scottish associationist tradition and was developed and transformed by physiological psychology during the nineteenth century. William Hamilton had argued that our sense of coherent selfhood is created by a process of 'latent mental modification' whereby we draw on a reservoir of unconscious mental traces, and he used a well known case, cited by Coleridge, in which a young German servant suddenly started speaking Hebrew and Greek to illustrate how all mental traces are stored in the obscure recesses of the mind (see Taylor and Shuttleworth, pp.75f.). 'Double consciousness' – initially defined as that condition where the mind automatically juggles manual skills or mental impressions – was considered to be rooted in the physical division of the brain by phrenologists in the 1830s and 1840s. But by the 1850s, these mental states were seen as the product of the memory, as always existing in time. Darwin's cousin, the physician Henry Holland, for example, compared conditions of total and partial dissociation of memory, 'where complete trains of thought are perverted and deranged, while others are preserved in sufficiently natural course to become a sort of watch upon the former' (Taylor and Shuttleworth, p.129).

By the 1870s, dramatic cases in France of *la dédoublement de la personnalité* – most famously Felida X and Louis Vivet, both cases of melodramatically alternating selves – led physicians such as Eugene Azam to conclude that the mind's ability to selectively recall past events could lead to the splitting not simply of consciousness, but of the personality itself, as the case study became established as the central explanatory method in neurophysiology. 'Sometimes the regretful longing turns into a bitter sense of exile, of banishment, of fall from high estate', noted Myers, describing these cases to an English readership in the *Nineteenth Century* in 1886. 'These instances have shown us the *retrogressive* change of personality, the dissolution into incoordinate elements of the polity of our being' (Myers, p.654). Such examples of selective amnesia contributed to the analysis of the psychological roots of bodily symptoms in cases of hysteria which would form the basis in the 1890s of Pierre Janet's theory that suppressed traumatic memories could become unconsciously fixed in the mind, and of Freud's argument that these suppressed memories would most likely be rooted in seduction during childhood (Hacking, pp.159-197; Micale and Lerner, pp.128-136; Sadoff, pp.28ff.). But the principal figure in the emergence of the sciences of memory in the 1870s and 1880s was Janet's predecessor at the Collège de France, Théodule Ribot. Ribot was well known in England both for his contribution to degenerationist discourse, *L'Hérédité Psychologique* (translated in 1875 as *Heredity: A Psychological Study of its Phenomena, Laws, Causes and Consequence*), which Gissing read and annotated in the original (Korg, 1962, pp.59ff.), and for the series that followed it, *Diseases of Memory* (1882), *The Diseases of Personality* (1885) and *The Diseases of the Will* (1883).

A staunch materialist, critical of the metaphysical interpretations of the unconscious will that Eduard von Hartmann's influential treatise *The Philosophy of the Unconscious* (1866) had adapted from Schopenhauer, Ribot in his account of 'organic memory' drew together a bewildering array of descriptions of the bizarre workings of the faculty – including Coleridge's anecdote of the German girl – and recast them in the light of Helmholz's analysis of energy conservation and dissipation (as David Glover shows here in his detailed discussion). However, to describe the actual workings of the mind Ribot drew on mid-century English physiological psychologists, particularly G.H. Lewes and Herbert Spencer, who had argued that reiterated ideas are incorporated into the brain through the nervous system, moving through subconscious habit, to unconscious reflex and hereditary instinct (see Otis, pp.14-25). Lewes had stressed that physiological explanations alone could not account for human developments. 'The mental life of man,' he maintained in *Problems of Life and Mind* in 1874, has 'a double root, a double history' that is social as much as animal, 'and no explanation of mental phenomena can be valid which does not account for this extension of range.' Spencer had argued that mental evolution takes the form of an 'antagonistic' process of integration and differentiation – on the one hand acquired habits grow increasingly unconscious, on the other the movement from simple to complex states of consciousness is manifested in the development of the will (see Taylor and Shuttleworth, pp.90, 86).

It is this model that Ribot develops through his work, and here his account of the relationship between conscious and unconscious suggestion inverts Schopenhauer's rather than absolutely opposing it. While Schopenhauer defined *das Will* as a force whose power resided in remaining unconscious, Ribot posited a hierarchy of consciousness in which volition is the 'last term of a progressive evolution', but which is ultimately determined by physiology, the seat of unconscious cerebration (Ribot, 1896, p.113). 'Every state of consciousness', he noted, 'represents only a very inconsiderable portion of our psychic life, because every instant it is supported and... impelled by unconscious states. Each volition... dives to the very depths of our being; the motives that accompany and apparently explain it, are never but a feeble part of its true cause' (Ribot, 1891, p.13). Both 'spontaneous acts of memory which appear to be incited by no association' and conditions of double consciousness thus 'throw light on the nature of the Ego and the conditions for the existence of a conscious personality' by dramatizing how 'the *Ego* subjectively considered consists of a sum of conscious states', Ribot proposed in *Diseases of Memory* (p.107). Godwin's recollection of the canon's sermon is a clear example of this kind of spontaneous memory driven by an impulse toward social assimilation; but the narrative immediately complicates this by focusing on the 'profound moral crisis' ('moral' is used here in the nineteenth-century sense which includes affect, or emotion, as well as moral judgement) of which his unconscious recollection is a symptom – on the 'double root, the double history' (in Lewes' terms) of his psychic condition.

Born in Exile thus reorders and develops Ribot's depth model by making Godwin's moment of spontaneous recollection both the product of and a radical departure from his inner history as it has developed through the novel so far. The narrative patterning also implicitly makes use of Ribot's complex model of identity as 'a sum of conscious states'. 'There is a central group surrounded by a secondary group which tends to supplant it,' Ribot argued, 'and these in turn are encompassed... with other less conscious states' in an 'unstable compound phenomenon' – all underpinned by 'the undefined consciousness, the product of all the vital processes constituting bodily perception', which together form the *coenaesthesis*, the sense of being at home within the world (Ribot, 1882, pp.98-108). Godwin's encounter with the Warricombes at Exeter triggers an extension of the explicit moment of double consciousness experienced at the Moxeys' ten years before. But now instead of experiencing two trains of thought he knowingly assumes a false persona as a response to what the narrator identifies as inextricably physical, psychic and social drives – a 'sum of conscious and less conscious states' that are rooted in Godwin's family history, are shaped and consolidated by his particular experience at Whitelaw College, and embody the tensions in social theory and psychology which converged on education during the 1870s and 1880s.

Education, mobility and the structures of consciousness

Throughout Gissing's fiction, education represents the possibility of individual transformation and marks the limit of the liberal ideal of progress though culture. These limits are probed most sympathetically in *Thyrza* (1887) and *A Life's Morning* (1878), while the destabilization of caste, the rise of sham 'culture', the nervous vulnerability of women and the exhaustion of modern life collide most dramatically in the assessment of mass education in *In the Year of Jubilee* (1894). However, *Born in Exile* is Gissing's fullest analysis of how an educational experience at once reinforces and disturbs established intellectual and social divisions, as Godwin's liminal social position produces an alienated psyche that disrupts the anomalies created by the continuing dominance of the leisured middle class as the preserve of 'traditional' culture in the very struggle to adapt to it. The detailed opening account of the prize-giving ceremony at Whitelaw College in 1874 (the end of Gissing's first year at Owens College, before his disastrous expulsion) dramatizes these tensions particularly starkly in the alternating perspectives of the historicizing narrative voice and the collective vision of the Warricombes.

For despite the extension of state-funded elementary schooling after 1870, education was still predominantly regarded as the means of shaping and policing class boundaries, while allowing a limited degree of social mobility. The reform of the endowed grammar schools following the Taunton Commission Report of 1868 had reinforced a classical liberal education at the upper end of a graded hierarchy of secondary schooling (Simon, pp.88-130). And while there was an increasing sense of the need of the middle classes to adapt to modern technological and bureaucratic challenges, the partially reformed ancient universities had quickly reinforced their dominance at the centre of national life by the late nineteenth century, drawing on the expanding ranks in commerce and the professions (Anderson, pp.17ff.). Gissing's own education reflected these tensions, as John Sloan has noted (Sloan, pp.3-9; see also Coustillas, 1970); but they were also manifest within Owens College itself. Founded in 1851 by the advanced Manchester liberal John Owens, it had followed University College, London (for which it prepared students for matriculation) in offering a broad curriculum free of religious tests, and it traced its genealogy to the Manchester College of Arts and

Sciences set up by the Literary and Philosophical Society in the late eighteenth century (Thompson, pp.1-44). Owens aimed to draw on the radical developments in science on the Continent and in Scotland, while servicing Manchester's diverse aspirations – together with the social and cultural divisions on which they rested – by offering an expanded Arts curriculum, professional training, and a separate strand of working-class education, incorporating the Working Men's College in 1861 (Jones, pp.65-94). The College's relaunch in new premises in 1870 was attended by such luminaries as T.H. Huxley, Sir John Lubbock and John Tyndall, and while the Professor of Greek gave a lecture on 'The relations of culture to practical life', the newly appointed Professor of Chemistry spoke on 'Original research as a means of education' (Thompson, pp.471f.).

Owens College therefore embodied some of the progressive educational ideals espoused by T.H. Huxley – whose 1868 article 'A liberal education and where to find it' Gissing had read and admired in March 1879 (Mattheisen et al., p.162) – while ultimately remaining trapped within the divisions that Huxley strove to overturn. Huxley had advocated 'the instruction of the intellect in the laws of Nature' alongside 'the facts of history, the process of thought, the conditions of moral and physical existence and laws of physical nature' (Huxley, p.83). His perspective embodies both the views attributed by Gissing to the collective body of Whitelaw when he calls them 'supporters of that mode of freedom which regards a life as a remorseless conflict' (*Born in Exile*, p.11), and the detached eye which observes these. In *Born in Exile*, Godwin rejects, then later returns to, the classics element of the Arts course he has studied at Whitelaw; but he also wins prizes in the natural sciences. Moreover, during the 1870s the Arts syllabus – praised by the *Revue Scientifique* – that Gissing actually studied included chemistry, natural philosophy and mathematics, as well as Greek and Latin, history and English literature (Owens College Calendar, 1873, cited in Farrelly, pp.20f.; Jones, pp.50ff.). Godwin's initial aim in going to Whitelaw was to study for the civil service examinations, and Owens too stressed the vocational diversity of its Arts programme, on which 70 per cent of students were from the professional and commercial middle class (Jones, p.153). In the explicit anxieties prompting Godwin's rejection of a classical education, and his return to it ten years later, we can detect how even though Owens placed science at the centre of its liberal education, the college's aim of heightening social polish overshadowed its progressive intellectual aspirations. This tension is embodied in the contrast between the two professors of science, Gale the genteel geologist and Walsh the more subversive *déclassé* chemist. Thus the syllabus incorporated science within the Arts curriculum while turning it into a gentlemanly pursuit rather than a radical intellectual challenge, not least by maintaining the institutional and social divisions between pure and applied science.

Whitelaw thus encapsulates the class and cultural duality – the institutional 'double consciousness' – that Earwaker successfully negotiates as a man of letters, but which only highlights the dissonance between Godwin's intellectual identity, his social ambition and his emotional drives. The opening prize-giving scene vividly dramatizes the intensely competitive ethos of secondary and higher education, which had lead to the anxieties about overwork and 'brain-forcing' used by some physiological psychologists who located education within a bodily economy and opposed the extension of education to lower middle-class men as much as women during the 1870s and 1880s. The eminent physician Sir James Crichton-Browne even warned against the expansion of state education for this reason, and emphasized that 'the evils of brain forcing are most likely to show themselves under a system of "cram" or spurt teaching, with a view to a specific examination, or of learning by rote and rule....' (see Taylor and Shuttleworth, pp.338f.; Oppenheim, pp.154-166, 239-247).

'A young man of spare figure and unhealthy complexion', with a 'forced, anxious demeanour', a 'bilious countenance', and 'overweighed with brains' (*Born in Exile*, p. 22), Godwin at first seems to embody such anxieties; he too spends time in church 'committing to memory lists of dates and the like, jotted down on a slip of paper' while cramming for the Whitelaw scholarship (p.44). But such 'spurt-learning' is primarily displaced onto those others who echo and undermine his own aspirations: onto Jowey Peak, with his rendering of 'The Ancient Mariner' in which 'the reciter stopped with the sudden jerk of a machine', and onto Bruno Chilvers, a 'rote-repeater of academic catch-words' (pp.80, 54f.). Godwin's sallowness is caused by hunger rather than hereditary weakness; his own physiology, combining nervous expenditure ('overweighed with brains') with assimilation and adaptability ('large brain, generous blood, delicate tissues'), is symptomatic of a far more perverse, contradictory sense of self. For as dangerous as educational over pressure, Crichton-Browne noted in 1895, were those moments of double consciousness, or 'dreamy mental states' often experienced by sensitive young men – 'revivals of hereditarily transmitted or acquired states in new or special combinations', which 'enfeebled' the 'highest nerve centres' and signalled mental dissolution (Crichton-Browne, pp.5, 39). In using such moments of double consciousness to highlight Godwin's ambiguous class identity, *Born in Exile* intersects with some of the wider debates on the shaping of subjectivity that surrounded education and evolution by making an explicit link to debates on memory, consciousness and the will.

In particular, there are some intriguing points of contact with J.M. Guyau's *Education and Heredity*, which appeared posthumously in English as Gissing was writing *Born in Exile* in 1891. Following Ribot, and echoing Crichton-Browne, Guyau stressed that the educational system 'should be oriented to the maintenance and progress of the race' by the inculcation of mental habits which would be passed to future generations; but his emphasis was on the limits of heredity, as he recast older environmentalist theories of moral management within an evolutionary framework (Guyau, p.xix). Guyau disagreed both with rigid hereditarians such as Galton, who argued that racial progress was made at the expense of extraordinary individuals, and with Ribot's earlier argument in *Heredity* that education can only gradually improve the average intellect and can have no impact on exceptional minds (pp.xxi, 105). Instead, he stressed, with Lewes and Huxley, that it was ultimately impossible to distinguish between the physiological and the social environment by insisting that the key to understanding the limits of individual adaptability and the possibility of willed transformation lay in the links between volition and unconscious impulses. In 'Multiplex Personality', F.W.H. Myers had emphasized the controlling and transformative power of hypnotism and Guyau, too, argued that children's suggestibility meant that like hypnotized subjects they could be shaped to higher social and moral ends (pp.2-32).

Guyau thus argues that the individual can only become a self-activating agent through understanding the power of external suggestion and the boundaries of self-control, and this means that he develops Ribot's analysis by stressing that physiology cannot be separated from either psychological or cultural meanings and effects, manifested performatively. 'Our social status', he notes, 'constantly suggests to us, in all circumstances, and often even in spite of hereditary tendencies, the conduct appropriate to that status' (p.39). Like Ribot, Guyau regarded conscious volition as the highest stage in mental evolution; but he reworks the implication of Ribot's emphasis on the divisions within the self. Like Ribot, he stresses that 'voluntary acts' are not the 'simple transformation of a detached state of consciousness [but] presuppose the participation of the whole group of conscious and subconscious states which constitutes the individual at any given moment'; but he concludes that 'the idea of a voluntary act is therefore by its very definition the idea-force which possesses most practical power in our consciousness' (p.64).

Guyau thus appropriates Ribot to shape a kind of materialist idealism, in which the multiplicity of the self, while rooted in physiology, is only realized through representation. 'The group of conscious and subconscious states that constitute the ego', he writes, 'is nothing but the shifting equilibrium of *representations of action*, to which corresponds an impulsive force, roughly proportional to the force of the representation itself. Our ego is but an approximation, a kind of permanent suggestion.' He continues:

> We shall never succeed in reducing to complete unity... all the systems of ideas and tendencies which are struggling within us for existence.... Those patients whose personality is doubled or even trebled show us, in an exaggerated form, the phenomenon constantly going on within us – the coexistence of several centres of attraction in our consciousness, of several currents crossing the field of consciousness, each of which currents, if not limited by another, would submerge us and carry us away.... In the depths of each of us there are more *selves* than one, whose shifting equilibrium constitutes what we imagine is our real *self*, which is in fact only our *past* self.... This ego is only ours in so far as our past determines our future; and there is nothing more variable than this determination of the future of a being by its past. It is true that our body serves us as a centre of reference; it is the basis of our personality. But the body is for us only a system of perceptions, and therefore of sensations, which, from a deeper point of view, reduce to a system of favoured or thwarted tendencies.... Our ego is therefore an *idea* ... which maintains our identity – though that identity is incessantly threatening to disappear into peculiar and present phenomena. (pp.64-66; original emphasis)

Guyau concluded that the aim of education should be to shape a continually evolving collective consciousness through the development of the 'idea-force'; but his analysis stresses the tenuousness of the isolated ego as a fiction that can only be created by an already overdetermined will and is only performatively effective within social roles.

Duality and its breakdown

In *Born in Exile*, double consciousness becomes a crucial means of exploring Godwin's ultimately unsuccessful struggle to shape a fiction of himself as both inside and outside a class-bound society through a rationalized act of will. In thus exploring the workings of his consciousness within an equivocal social environment – including others' view of him – the narrative extends contemporary debates on memory and identity. An unstable trope, mental duality enables dissonant meanings to be held together within an embedded structure by a voice that shifts between identification and distance in its simultaneous analysis and representation of a consciousness that teeters on the brink of control of its own associations. To conclude I'll briefly discuss how this works through the two countervailing drives towards separation and assimilation in Godwin's mental life, exploring how Gissing reframes the implications of particular models of suspended or altered consciousness in particular contexts.

'With the growth of his militant egoism, there had developed in Godwin Peak an excess of nervous sensibility which threatened to deprive his character of the initiative rightly belonging to it...' (*Born in Exile*, p.54). Noting that Godwin has lost much of 'his natural vigour which declared itself at Andrew Peak's second visit to Twybridge' after the prize distribution at Whitelaw, the narrator sets up the self-cancelling pattern that will be elaborated throughout *Born in Exile* as Peak moves through a double process of differentiation and integration. Godwin's 'natural vigour' is manifested, in Ribot's terms, as a kind of *coenaesthesis* – the 'undefined consciousness' rooted in the body: it is manifested, for example, in the 'rush of confident blood, pulsing through all the mechanism of his being' that Godwin experiences during his long trudge from Kingsmill to Twybridge as he ponders his 'noteworthy powers' (p. 61). However this is immediately followed by the moment of lapsed consciousness in which he experiences the vastness of geological time; and behind it lies the complex history that is elaborated through the series of ellipses in the first part of the novel. Here for the first time Gissing implicitly dramatizes Guyau's argument that the body is 'only a system of perceptions' and that 'there is nothing more variable than this determination of a future of a being by its past', as the moment in 1874 (which forms the immediate past of the rest of the novel) is bifurcated by the flashback to Godwin's family history and childhood. Again, this forms a 'double root, a double history' which shapes both Godwin's ambiguous class position and his ambivalent social and intellectual aspirations, and forms a critical perspective from which to view his own self-representations.

For Peak himself often negotiates and rationalizes his situation, as either determined by his family history or transcending it, through a selective reading of Ribot's *Heredity*; and this paradoxically serves to highlight how profoundly he is moulded by circumstance. ' "Do you know what is meant by inherited tendencies?" ', he asks his brother Oliver. '... "Children don't always take after their parents; very often they show a much stronger likeness to a grandparent or an uncle.... Read, read, read!" ', he admonishes, echoing Gissing's own note on Ribot's text, that 'the individual ceaselessly tends to modify himself' (pp.65f; Korg, 1962, p.61). Godwin regards himself, in Ribot's terms, as one of those exceptional minds who can transcend their heredity, compared to his contemporaries – dismissed as 'the stuff which destiny uses for its everyday purposes, to keep the world a-rolling' (*Born in Exile*, p.67). Yet his cousin Jowey reflects his own aspirations, and together with his paternal and maternal inheritance – his father's radical scepticism, his mother and aunt's snobbish upward mobility – this family network manifests the multiple forms the drive for individual improvement can take. It places Godwin firmly within the lower middle-class world that he attempts to transcend, even as he regards his siblings as having 'relapsed, instead of bettering the rank their father had bequeathed to them' (p.127). Stephen Ogden has suggested that Godwin's implicit application of the principles of sexual selection represents Gissing's own growing unease with Darwinism (Ogden, pp.171ff.). But Godwin's reification of Sidwell and his 'artificial' reinforcement of Martin Warricombe's outmoded beliefs interrupt the Warricombe family's psychological evolution in Ribot's terms. Peak's approval of the very hereditary class structure which he disturbs bears out and is ironised by Ribot's own comment in *Heredity* on the stagnancy of caste societies (Ribot, 1875, p.364).

Godwin's assimilation of the moment of lapsed consciousness in which he performatively enacts a role that he will later knowingly adopt complicates this already intricate concept of heredity as unconscious memory. His wish to construct such a position and identity for himself dramatizes the process that both Ribot and Guyau describe, whereby a volition is prefigured by a subconscious impulse; but this transmutes, in Guyau's terms, into a representation created by the 'idea force' of his will. Godwin attempts to construct a fiction of a coherent identity by rationalising his position to himself, on the one hand by justifying religion socially as a useful containing force, on the other by abstracting Sidwell into an idea. 'So strangely had circumstances moulded him', remarks the narrator, 'that he could not brood on a desire of spiritual affinities... [He] was preoccupied with the contemplation of qualities which characterize a class' (*Born in Exile*, p.218). And this means that the deliberate 'cultivation of a double consciousness' that Buckland Warricombe (exposing the class-bound limitations of his own radicalism) interprets as illegitimate imposture; that Earwaker, more subtly, reads as the assumption of an invented persona; and that Malkin farcically amplifies by seeing Peak's '*doppelganger*' in New York, is inexorably undercut by particular modes of disordered association.

For the tension between Godwin's sceptical and created self which grows out of his own self-consciousness often takes the form of that mode of double consciousness in which alternative selves are shaped by divisions within the strands of memory. This is most pronounced at the moment when he is giving the fullest rendering of the intellectual position that he rejects, as when he provides Martin with a vulgarised evolutionary interpretation of Christian teleology:

> Peak had been talking for more than a quarter of an hour. Under stress of shame and self-criticism (for he could not help confuting every position as he stated it) his mind often wandered. When he ceased speaking there came upon him an uncomfortable dreaminess which he had already once or twice experienced when in colloquy with Mr Warricombe; a tormenting metaphysical doubt about his own identity strangely beset him. With involuntary attempt to recover the familiar self he grasped his own

> wrist, and then before he was aware, a laugh escaped him, an all but mocking laugh, unsuitable enough to the spirit of the moment. (p. 252)

This passage echoes the first episode of double consciousness at the Moxeys in which Godwin attempts to carry on two strands of thought simultaneously; but here the direct conflict between his dual perception and his struggle to suppress his sceptical self paradoxically gives his own rationality the simultaneously somatic and psychic power of an unconscious impulse. His attempt to 'recover his familiar self' by grasping his own wrist places his body at the centre of a process in which his 'real self' – or in Guyau's terms, his past self – becomes as tenuous as his false persona. Identity eludes bodily *coenaesthesis* even as his 'militant egoism' is undermined by the 'excess of nervous sensibility' which is its inevitable concomitant.

This means that Peak's 'cultivation of a double consciousness' is inexorably undermined by the very psychic structures that, in a different register, it mimics, as sexual desire, emotional hunger and social ambition are condensed into a boundless need expressed though imaginary projections and shadowy representations. Godwin, Mrs Peak observes, is like his father: ' "There was always something he wanted, and yet he didn't know what it was" ' (p.437). As he moves towards extinction as a social anomaly whose psychic identity depends on the very conflicts he at once embodies and struggles to transcend, that something fleetingly becomes the self he might have become as the narrative, too, finally detaches itself from him and recedes. 'Godwin Peak is an interesting and genuine study in human nature, and we cannot lightly dismiss him as a snob', commented *The Times*, reviewing the novel in July 1892. '.... However it could hardly have failed to occur to Peak that in the nineteenth century he was far more likely to attain the social position he coveted by persevering with his own work than by masquerading as a clergyman' (Coustillas and Partridge, pp.203f.). But it is precisely the half-acknowledged perception of this possibility, again rooted, indirectly, in his time at Whitelaw College, that shapes Godwin's mental conflicts in the final part of the novel as the structures of double consciousness break down, together with the performative persona that they have sustained.

For while Earwaker and Chilvers represent the career trajectories produced by the Arts education at Whitelaw that Godwin is unable to reconcile (and can only attempt to negotiate by acting as if none of his fellow-students have any memory), the possibility of an integrated intellectual and social identity is nonetheless briefly introduced, if only to be snatched away. Peak's final possible alter ego is not Chilvers or Earwaker, but his brief rival for Sidwell's affection, Justin Walsh, brother of that 'object of his rebellious admiration' the 'ungentlemanly' Whitelaw professor of chemistry. A 'modern of the moderns; at once man of science and man of letters; defiant without a hint of popular cynicism, scornful of English reticences yet never gross', Walsh represents the rejected future that Godwin acknowledges: he 'felt his nerves thrill in sympathetic response... Instead of mouldering in a dull cathedral town he might now be living and working in France or Germany' (pp.345f.). Even as an object of homosocial exchange, Sidwell acts as a tacit link between Godwin and Walsh; and her own identity becomes increasingly liminal, in a way that suggests Guyau's concept of intellectual sympathy, as she absorbs Godwin's scepticism. And while Sidwell finally withdraws back into the Warricombe family, Christian Moxey succeeds in casting off his fantasies of femininity and marries his cousin Janet, the doctor who has successfully broken down the barriers to women's higher education (albeit paying the price of overwork).

Godwin's own 'inner voice' thus both suggests the self that he might have become and acts out the patterns of emotional ambivalence that make this alternative identity impossible. However it does so by taking the form of irony, rather than the 'inherited disease' of conscience (p.441). It is irony, the frisson produced by the clash of voices, that ultimately becomes the 'double consciousness' which not only expresses Godwin's sceptical self but causes him to teeter on the brink of self-control through an excess of self-consciousness – an excess that is almost pathologised but not quite, as the narrative moves between comment and identification. 'The ironic temptation was terribly strong in him just now', the narrator notes at the moment when Peak has been discovered by Marcella Moxey and his two selves threaten to come face to face. 'One is occasionally possessed by a desire to shout in the midst of a silent assembly; an impulse of the same kind kept urging him to utter words which would irretrievably ruin his prospects' (pp.306f.; Myers, p.649). 'We have nothing to do with names and epithets', Godwin tells Earwaker in a moment of self-justification – but such special pleading is immediately undercut by the interjection of this other self: 'Since his farewell interview with Sidwell he knew no rest from the torment of a mocking voice which bade him bear in mind that all this dishonour had been superfluous...' (*Born in Exile*, p. 441).

'We shall never succeed in reducing to complete unity... all the systems of ideas and tendencies which are struggling within us for existence', Guyau noted, and Godwin is finally 'submerged and carried away' as the material conditions that determined his cultivation of a double consciousness are lifted and he is free to become his 'real self'. *Born in Exile* ultimately refuses to resolve the contradictory elements of Peak's identity: as Godwin sets out on his wanderings in Europe, towards death in exile, he sends Earwaker his photograph, an image of 'a face by which every intelligent eye must be arrested; which no two observers would interpret in the same way' (p.483; and see Sloan, p.114). But just as the photograph captures his image without offering any stable interpretation of it, so the narrative makes use of the many-faceted discourse of double consciousness to write its symptomatic history of a man who remains a strange, indecipherable case, and who pushes the radical implications of late nineteenth-century discussion of memory, identity and the will to their limits, and beyond.

Bibliography

Alden, Patricia (1986), *Social Mobility in the English Bildungsroman: Gissing, Hardy, Bennett and Lawrence*, UMI Research Press, Ann Arbor and London.

Anderson, R.D. (1992), *Universities and Elites in Britain since 1800*,, Macmillan, London.

Coustillas, P. (1970), *George Gissing at Alderley Edge*, Enitharmon Press, London.

Coustillas, P., and Partridge, C. (eds.) (1972), *Gissing: The Critical Heritage*, Routledge and Kegan Paul, London.

Coustillas, P. (ed.) (1978), *London and the Life of Literature in Late Victorian England: The Diary of George Gissing, Novelist*, Harvester Press, Brighton.
Crichton-Browne, James, (1927; first publ. 1895), 'Dreamy mental states', *Stray Leaves from a Physician's Portfolio*, London.
Farrelly, Carol (2003), 'Nature's aristocrats? Education and social mobility in *Born in Exile* and *Jude the Obscure*', unpublished paper, University of Sussex.
Gissing, George (1985; first publ. 1892), *Born in Exile*, The Hogarth Press, London.
Goode, John (1978), *George Gissing: Ideology and Fiction*, Vision Press, London.
Guyau, J.M. (1891), *Education and Heredity. A Study in Sociology*, Walter Scott, London.
Hacking, Ian (1995), *Rewriting the Soul: Multiple Personality and the Sciences of Memory*, Princeton University Press, Princeton.
Hardy, Thomas (1987; first publ. 1895), *Jude the Obscure*, Penguin, Harmondsworth.
Huxley, T.H., 'A liberal education and where to find it' (1893; first publ. 1868), in *Essays on Science and Education*, Macmillan, London.
Jones, David, R. (1988), *The Origins of the Civic Universities*, Routledge, London.
Korg, Jacob (1963), *George Gissing, A Critical Biography*, Harvester Press, Brighton.
Korg, Jacob ed. (1962), *George Gissing's Commonplace Book*, New York Public Library, New York.
Mattheisen, Paul F., Young, Arthur C. and Coustillas, Pierre (eds.) (1990), *The Collected Letters of George Gissing*, vol. I, Ohio University Press, Athens, Ohio.
Micale, Mark S. and Lerner, Paul (eds.) (2001), *Traumatic Pasts: History, Psychiatry and Trauma in the Modern Age, 1870-1930*, Cambridge University Press, Cambridge.
Myers, F.W.H. (1886), 'Multiplex Personality', *The Nineteenth Century*, 20, pp.648- 666.
Ogden, Stephen (2001), 'Darwinian Scepticism in George Gissing's *Born in Exile*', in Bouwe Postmus (ed.), *A Garland for Gissing*, Rodopi, Amsterdam, pp.171-178.
Oppenheim, Janet (1991), *"Shattered Nerves": Doctors, Patients and Depression in Victorian England*, Oxford University Press, Oxford.
Otis, Laura (1994), *Organic Memory: History and the Body in the Late Nineteenth and Early Twentieth Centuries*, University of Nebraska Press, Lincoln and London.
Ribot, T.A. (1875), *Heredity: A Psychological Study of its Phenomena, Laws, Causes and Consequences*, Henry S. King, London.
Ribot, T.A. (1882), *Diseases of Memory*, Kegan Paul, Trench and Co, London.
Ribot, T.A. (second English edition, 1896; first publ. 1884), *The Diseases of the Will*, Open Court Publishing, Chicago.
Ribot, T.A. (second English edition 1891; first publ. 1883), *The Diseases of Personality*, Open Court Publishing Company, Chicago.
Sadoff, Dianne F. (1998), *Sciences of the Flesh: Representing Body and Subject in Psychoanalysis*, Stanford University Press, Stanford.
Simon, Brian (1987), 'Systematisation and Segmentation in Education: The Case of England', in Muller, D.K., Ringer, F. and Simon, B. (eds.), *The Rise of the Modern Educational System: Structural Change and Social Reproduction*, Cambridge University Press, Cambridge.
Sloan, John (1989), *George Gissing, The Cultural Challenge*, Macmillan, London.
Swann, Charles (1984), 'Sincerity and Authenticity: The Problem of Identity in *Born in Exile*', *Literature and* History, 10: 2, pp.165-188.
Taylor, Jenny Bourne and Shuttleworth, Sally (eds.) (1998), *Embodied Selves: An Anthology of Psychological Texts 1830-1890*, Oxford University Press, Oxford.
Thompson, Joseph (1880), *The Owens College: Its Foundation and Growth and Its Connection with the Victoria University, Manchester*, J.E. Cornish, Manchester.

Chapter 5

Sex and the City: Gissing, Helmholtz, Freud

David Glover

The realization that George Gissing and Sigmund Freud were writing and publishing in the same historical moment still comes as something of a shock. After all, there could scarcely be two more different figures: the brilliant but thoroughly disgraced scholar who despite his aborted education was to become England's nearest approximation to a Naturalist novelist; the ambitious young Viennese doctor who revolutionized the investigation of psychic life, leaving suspicion and scepticism in his wake. What, other than sheer coincidence and a breath of scandal, could possibly align them?

In seeking an answer to that question, this essay attempts to read Gissing with Freud historically, charting a network of sources in medico-scientific thinking on which both authors drew, though to different effect. While its primary emphasis is upon Gissing, it also focuses upon the 1890s as an especially fruitful period for both writers, and it finds a common influence upon their writing in the work of the scientist Hermann von Helmholtz. Though of course psychoanalysis plays no role in Gissing's novels, this was the decade that witnessed its birth. With the publication, in 1895, of Freud's groundbreaking *Studies on Hysteria*, co-authored with Josef Breuer, psychoanalysis effectively came of age; and a year later, in an allusion to this work in an article on 'L'hérédité et l'étiologie des névroses', Freud first referred to his 'new psychoanalytic method' by name (Freud, 1963, p.145). But Gissing's admittedly more advanced career also saw a new point of departure in the 1890s. His writing began to move away from the lower reaches of metropolitan life and turned instead towards the insecurities of the professional and sub-professional middle classes – precisely those strata that formed the mainstay of Freud's emergent clientele. Commencing with his pitiless anatomy of the literary marketplace in *New Grub Street* (1891) and continuing through *The Odd Women* (1893), *In the Year of Jubilee* (1894), *Eve's Ransom* (1895) and *The Whirlpool* (1897), Gissing produced a remarkable series of novels that probed the faltering fortunes and precarious inner worlds of those whose claims on cultural capital were never quite strong enough to insulate them from the vicissitudes of an increasingly rapacious commercial environment. Gissing had always specialized in blighted lives: now he found an unexpected measure of success in charting the irresistible pressures of late Victorian modernity. Part of his appeal surely lay in the fact that few of his readers found themselves wholly untouched by such currents – the growth of advertising, the industrialization of leisure, the rise of a new reading public, the impact of the joint-stock limited liability company – just as today's readers continue to flounder in their backwash.

In the Year of Jubilee and the 'facts of physiology'

If Gissing's writing from the nineties registers some of the major social and economic developments of the period, it is the manner in which these trends were refracted through the sphere of intimate personal relations that distinguishes his later work. Believing that 'lives are ruled by [the] minutest facts of physiology', as he once confided in a working notebook, Gissing increasingly sought to incorporate a wider scientific rationale for this maxim into his fiction (Korg, p.61). The peculiar combination of abstraction and particularity that results from accentuating the problem of the explanation of human behaviour is what gives a novel like *In the Year of Jubilee*, in which private dramas are conspicuously linked to a mass public event, its unusual cautionary or despondent tone. Moreover, the ultimate lure of such narratives is typically, as it has so often been for psychoanalysis, that of the woman: her resistances, her hysteria, her symptoms, her femininity. And it is here that one finds a significant line of convergence between Gissing and Freud. As Rachel Bowlby has noted in an unflinching comparison, the two writers used much the same language in their discussion of women, a vocabulary in which femininity can only be a problem or (what would undoubtedly be the same thing) a provocation (Bowlby, p.42). In the novella *Eve's Ransom* (1895), for example, Maurice Hilliard's search for the real woman whose photograph in his landlady's album 'doesn't do her justice' is transformed into an attempt to reclaim her compromised sexuality from the impasse into which it has fallen, an act of erotic salvage that is in turn sublimated by ceding her to another man (*Eve's Ransom*, p.17). Hilliard retains and reinforces his mastery over the eponymous Eve by letting go of her, by allowing her to pass between men, by exchanging his sexual investment in her for a renewal of his independence, albeit with considerable emotional difficulty. He is both more and less than the 'physician' whose role he at one point mimics with mock solemnity: too besotted to surrender his lover graciously, yet so calculatingly unstinting in his generosity that she can tell him ' "All I am and have I owe to you" ' (pp.66, 124). In any case, as Bowlby shrewdly observes, this is a 'cure' that effectively inverts the normal Freudian contract with the analysand insofar as it is the male analyst-substitute who in this instance offers money to his patient – though it is worth adding that she initially pays him back with what he regards as 'mere hysterical conscientiousness' (p.120). In short, the secret of femininity evidently has a literal as well as a metaphorical price – and, as a corollary, hysteria and breakdown start to move into the foreground at this point. One could almost say that they structure the terms of the sexual or affective bargain to which Gissing's fiction so obsessively returns throughout this period.

To provide a suitable setting for such fraught transactions between men and women – indeed, to find the theoretical means with which to conceptualize their difficulty – Gissing frequently skips back a decade and returns to the 1880s. It is as though this decade offered a kind of laboratory for understanding the *social*, as well as the psychological or biological roots of the dislocations of gender and domesticity, an understanding that paradoxically moves him both closer towards and further away from Freudian parallels. Thus, although *The Odd Women* begins with the sudden death of the lower-caste country doctor Elkanah Madden in 1872, his tragic end is no more than the premise for an examination of displacement and stunted opportunities that is deferred until 1887. Once the novel has properly begun it is immediately clear that the doctor's complacent assertion that ' "Human beings are not destined to struggle for ever like beasts of prey" ' forms the bitter epigraph to his orphaned daughters' subsequent experience of the 'coarse actualities' that characterize the nation's most advanced urban centre (*Odd Women*, p.32). Thrown back upon their own meagre resources, the Madden sisters subsist on the bare margins of impoverished respectability, hedged about by prostitution, alcoholism and grossly underpaid work.

What Gissing captures here and elsewhere, more impressively than any of his rivals, is the arrhythmia of modernity, its unevenness and stark discontinuities. Even in the elated grip of the Jubilee fever that inspired his 1894 novel, Gissing's London is shadowed by a threat of violence ready to erupt upon the thronging streets, as though the boisterous carnival atmosphere somehow carried within it the seeds of the window-smashing and looting that had been among the hallmarks of proletarian demonstrations in the West End in 1886 and 1887. The rediscovered signs of poverty evident from the early 1880s appear cheek by jowl here with an overall rise in real incomes and the increasingly aggressive rhetoric employed by English advertising, a juxtaposition that produces an uneasy mélange of social legitimations and self-interpretations. No one who follows the traces of these phenomena that are threaded into Gissing's *In the Year of Jubilee* can have any doubt that this is a novel that deals with a cultural and political watershed, however obliquely.

Although strictly speaking *In the Year of Jubilee* was the last triple-decker that Gissing was to publish, it in fact forms a kind of diptych with his next full-length novel *The Whirlpool*. For both texts hold up to critical scrutiny the very notion of civilization and, once weighed in the balance, find it to be severely wanting. Of course, this unrelieved pessimism, the slide from *fin de siècle* to *fin du globe*, was characteristic of the literary sensibility of the 1890s. But, while the idea that civilization's own processes of development were working against it was of much older provenance, one can see a refurbished version of this view beginning to be disseminated in popular scientific tracts in England in the mid-1880s, particularly as a way of comprehending the dangers of urban living. By 1888 it was possible for *The Lancet* not simply to ask 'Are We Degenerating Physically?' but to be sure that the answer was 'undoubtedly', listing such likely causes as 'sexual indulgence in early life, premature marriages, over-pressure in education', poor diet and 'increased tension' among 'town-bred populations' (Pick, p.201). Indeed, as Gareth Stedman Jones argued in his classic study *Outcast London* (1971), the various theories of degeneration that circulated among the professional middle classes provided a medium through which their fears for the future of the capital could be raised, given shape, and occasionally exorcized. As a well-informed and curious reader with a speculative turn of mind, Gissing was no exception to this ideological fashion, including Sir Francis Galton and Max Nordau among the authors he tackled while working on *The Whirlpool* and taking up William Hirsch's *Genius and Degeneration* (1896) shortly after finishing the proofs (Coustillas, pp. 365, 422, 434). In this regard, the dispositions of Gissing's characters are quite true to form. Mrs. Abbott, one of the novel's minor figures, is shown dutifully combing the French degenerationist Théodule Ribot's *L'Hérédité Psychologique* (1875) in an effort to explain the strengths and weaknesses of her family as well as those around her. This was a text that Gissing knew well, having written extensive notes on it in his commonplace book (Korg, pp. 59-62). While this citation would have had a definite resonance for readers in the late 1890s, its appearance in the novel was not at all anachronistic since *The Whirlpool* opens in London in 1886 at the precise moment when theories of urban degeneration were starting to gain a hearing. Indeed, the book's 'hero', Harvey Rolfe, also thoughtfully peruses studies of heredity at one point (p.136).

John Goode has claimed that *In the Year of Jubilee* inaugurated 'a new phase' in Gissing's writing, anticipating the more fully 'achieved' success of *The Whirlpool* (Goode, p.181). Certainly Gissing was very concerned to distance the former text from any association with the Woman Question, fearing that that his work might become typecast after *The Odd Women* just as public interest in feminism was waning. However, in some respects *In the Year of Jubilee* constitutes less of a break with his earlier writing than might be imagined. Gissing had long been inclined to see social class in essentially biological terms and both *Demos* (1886) and *The Nether World* (1889), for example, portray failure of character as a narrative of quasi-racial degeneration among the working classes. By contrast, *In the Year of Jubilee* deals with the worm in the middle-class bud. The paterfamilias Stephen Lord seems at first glance to be as 'forcible and commanding' as any 'substantial man of business', but his 'countenance' has 'a despondent cast' and 'the lips suggested a sullen pride, indicative of some defect or vice of will' (*Jubilee*, p.25). Worse still, these 'unpleasant characteristics' can already be identified in his favourite child Horace who not only has a 'frail constitution', but is already recognizably 'slow of intellect and weakly self-willed' (p.29). As this inherited flaw implies, neither man is able to avoid a disastrous marriage. Stephen Lord turns out to have been entangled with a woman who abruptly leaves and then divorces him, while he pretends to his children that she has died long ago. And, as if to continue on the same downward slope, Horace also marries a thoroughly unsuitable woman, the wayward and sexually precocious Fanny French whose death from consumption he finally counts as a happy release, shortly before his own. After this brief tormented childless union, the male line comes to an appropriately sterile end, its degeneracy expunged.

This is by no means the sole reference to putatively scientific thinking in the novel, for the references to degeneration flow back, via notions of the survival of the fittest, to the question of the relationship between heredity and nervous collapse. One of Gissing's more scathing ironies is to have his heroine Nancy Lord borrow a 'new book on Evolution' from her local library 'just before Jubilee Day', only to abandon it because it told her nothing she did not already know (p.97). Nancy associates the overarching Darwinian narrative with the sententiousness of her suitor, the autodidact Samuel Barmby – never happier than when he is lecturing 'to a society of mutual improvement' on a subject like 'National Greatness: its Obligations and its Dangers' (p.306) – and as a result she impatiently dismisses the volume as having little relevance to her own life. Smugly convinced that she could easily talk about Darwinism 'oh, with an air', she is quite unprepared for 'the battle of life' that overtakes her later in the novel, setting her cherished '"education" and her "culture"' at nought (p.297). Wandering the London streets in search of employment – and the sexual connotations of her fallen plight are unmistakeable, given that she has been deeply compromised through having rushed into a secret marriage barely a day after her lover Lionel Tarrant first carried her 'away into the shadow of the trees' (p.133) – Nancy imagines that '[t]he roar of myriad industries made mocking laughter at' her 'futile pretensions' (p.297). Even a humble seamstress or a matchbox-girl can command a higher market value in this social Darwinist universe.

When Nancy puts aside her library book, it is to the excitement of the Jubilee night celebrations that her mind irresistibly strays. In fantasy the movement of 'the huge crowd' acquires a distinctly masculine coloration partly through its association with the banter of the advertising impresario Luckworth Crewe whom she meets on the street and who helps to produce an experience that she feels 'was worth whole oceans of "culture"' (p.98). In fact, Nancy's sense of being caught up in the human flow is not so much oceanic as sexual. Adrift in what Gissing describes as the foolishly contented bestiality of the herd, she loses her reason, her 'identity' and her individuality in an emotional free-fall that robs her of her 'culture' and demotes her to the level of a common 'shop-girl'. Her bodily sensations are intensified once '[h]er blood was heated by close air and physical contact' and her social being has 'evanesced in this atmosphere of exhalations'. But from the outside, Gissing's omniscient voice insists scornfully, yet not without a touch of fascination, her face registers a 'look of vulgar abandonment', a look whose vulgarity would shock her profoundly (pp.68f.). The terms of Nancy's subsequent loss of respectability, her brush with sexual and economic degradation, are already being adumbrated.

Crewe is himself too irredeemably brash and vulgar for Nancy ever to seriously entertain him as a suitor – her ' "culture" ' still counts – but what he accomplishes is to provide the masculine hand that skilfully guides her through what can be read as a moment of orgasmic release. And he does this not merely through the medium of talk or chat, but by introducing a rudimentary scientific analysis of the dynamics of the crowd based upon his memories of 'physics at school'. Taking 'the resultant of forces' as his theoretical warrant, Crewe conceives of Nancy and himself as 'a force tending to the right' pitted against 'the crowd [that] is a force making for straight on' and although their makeshift efforts end 'in laughter' they do manage to reach a point from which 'walking was unimpeded' (p.72). It is therefore entirely apt that Nancy's next serious choice of library books should include Hermann von Helmholtz's *Popular Lectures on Scientific Subjects* (1873) with its references to the 'Conservation of Force' or the 'Interaction of Natural Forces' (p.121). Ridiculed as wholly fanciful by Lionel Tarrant, her husband-to-be, Helmholtz's physics and physiology of energy was in reality, alongside Darwinism, one of the great nineteenth-century scientific materialisms (painstakingly reconstructed by Anson Rabinbach in his important 1992 study *The Human Motor*). Out of Helmholtz's lectures 'emerged a new image of society powered by natural forces, of labour perpetually renewed, of the unity of society and nature in the process of mechanical – or industrial – production, and ultimately of the universe as a vast productive machine' (Rabinbach, p.59). What Helmholtz's ambitious research programme licensed was more than Crewe's mastery of a crowded thoroughfare: it conjured up the entire political economy of labour power from which Nancy Lord would eventually find herself excluded.

There was another, less expansive side to Helmholtz's system, however. By the 1860s he had come to believe – at least temporarily – that, as the second law of thermodynamics showed, energy suffered from a long-term propensity to decline or exhaust itself, to tend towards a condition of entropy. The universe was doomed to an irrevocable process of depletion, a sort of naturalized end-game. This near-metaphysical pathos added further scientific impetus to the widespread presumption that the natural and social world was perceptibly deteriorating, as theories of degeneration, for example, were also claiming, though from a somewhat different conceptual and empirical standpoint. These predictions began to converge in certain key instances. Thus, Théodule Ribot, whose work figures briefly in *The Whirlpool*, attempted to develop a Helmholtzian psychology of the will in which the latter was both defined as nature's 'highest force' *and* seen as chronically prone to dissipation or fatigue. In this search within physiology and psychology for what Anson Rabinbach has aptly called 'a science of fatigue', neurasthenia (that is, 'moral and psychic exhaustion') came to figure among the main diagnostic categories in mental pathology or nervous disorder, alongside hysteria, with which it was frequently associated (Rabinbach, p.67). Inevitably, therefore, Helmholtz, and the work he inspired, formed part of the universe of clinical ideas upon which Freud drew, although Freud's account of hysteria is distinguished as much by his sense of the limitations of Helmholtz's thought as by his belief in its utility (Sulloway, pp.65f.). What Freud questions more thoroughly than any of his predecessors is precisely the role of the body as the seat of mental disturbance.

Gissing had no such reservations, though he did recognize that the 'transmission' of inherited traits was 'a difficult point' because it so often seemed to be arbitrary and could be interrupted by 'accidental causes' (Korg, pp. 60f.). Consequently his attitude to the diagnosis of neurasthenia appears deeply divided (Greenslade 1994, p.149). One can see this ambivalence in his treatment of the sexually delinquent Nancy Lord. For all her troubles, Nancy is not offered up as the classic exemplification of the neurasthenic woman. By and large she evades the burden of the Lords' hereditary taint, despite her naivety, her impulsiveness and her shallowness. The implicit condemnation of her audacity is effectively displaced on to her plain and earnest confidante Jessica Morgan, a young woman who takes 'culture' so seriously that she wears herself out studying with 'hysteric determination' for a London University degree. Her ambition is to rise above the downwardly mobile family to which she belongs, to become 'one of the clever, the uncommon women'. Yet this exercise of her will is remorselessly exposed as nothing more than the crudest form of sublimation. The summary inventory of her unhealthy body amounts to a mere 'ghost of girlhood, a dolorous image of frustrate sex', her brain crammed with 'heterogeneous rubbish' learned by rote (pp. 16f.). Predictably the final examination occasions the return of her repressed desire for Samuel Barmby and, haunted by his face, she escapes into nervous collapse ('a most interesting case of breakdown from undue mental exertion') and fails her degree miserably (p.266). Abject and spiteful, Jessica ultimately channels her wasting sexual energy into religious evangelism which allows her to beg Nancy's forgiveness for having betrayed the secret of her clandestine marriage to Lionel Tarrant while simultaneously denouncing her erstwhile friend's sinful nature. Her parting gift to Nancy is 'a smile of rancorous compassion' (p.440).

The heavily caricatured figure of Jessica Morgan functions as a device that allows Gissing to highlight Nancy's virtues and to offset her vices. Nancy may be vain and headstrong, she may represent, in the words of Gissing's draft title, *A Girl's Wild Oats*, but she can never be as inept, as humourless, or as deeply unattractive as Jessica who signifies the *reductio ad absurdum* of New Womanly aspiration, the odd woman as an oddity pure and simple. Compared to Jessica's helpless lapse into hysteria, Nancy's impetuous failings and agonizing adjustments appear positively enlightened and independent-minded, as indeed they were calculated to do. Whereas Nancy is capable of real development, Jessica must perform an ever-diminishing cycle of repetitions, each cycle bringing her to a new low. Unlike Nancy's ability to transcend the narrow 'ancestral prejudices' of her family, class and neighbourhood (that 'provincialism' Lionel dubs 'Camberwellism'), Jessica Morgan's self-defeating desire for respectability finally pulls her down below the level reached by her already 'fallen' family (pp. 182f.). Her solution is to retrieve a sense of superiority by adopting a moralistic position that stands outside and above the 'sheer struggle for subsistence' in which her parents and siblings are mired (p.16). But in reality her salvation is parasitic upon the continued suffering of others.

In line with the prevailing medical orthodoxies of his day, Gissing does not distinguish clearly between neurasthenia and hysteria – though their differentiation is a major concern of Freud's contribution to *Studies on Hysteria* (Oppenheim, p.96). Thus, Jessica's condition combines different symptomatologies, muddying the line between body and psyche. Yet her illness is never narrowly physical or psychological. If the seeds of her breakdown can be found in her family's genteel poverty, this poverty is embedded in a wider system of predatory socio-economic relations. One instance that comes particularly close to home is Jessica's parents' jerry-built house, designed to flatter and gull the impecunious middle classes but so poorly constructed that it is almost impossible to inhabit with any comfort. And a similar tale of exploitation underwrites her father's partnership in a debt-collecting agency formed with 'an acquaintance' who provides 'the modest capital' necessary to launch the venture, while Morgan is cast as the reluctant 'bird of prey', 'straining his wits to circumvent the fraudulent, and swooping relentlessly upon the victims of misfortune'. Although he cannot understand why keeping up 'genteel appearances' is the mainspring of his life, Morgan does have the wit to recognize, with considerable distaste, that 'society' is 'a mere system of legalised rapine' (pp.

219f.). Against this background, Jessica is among the casualties of a viciously competitive and debased social order, her failure and breakdown as symptomatic of the overall shape of things as her father's gentlemanly aversion to his role of creditors' cats-paw. In general terms, the Morgans represent the cultural dead-end that Gissing deprecates so profoundly, and as such their fallibilities provide a satirical foil to the moral elements of the text privileged in the *Bildungsroman* of Nancy and Lionel. Nevertheless, despite all the heavy-handed parody and anti-feminist invective, Jessica's hysteria does emerge as a kind of cry of protest against a heartless world.

'Round and round they go'

To register this critical note involves reading *In the Year of Jubilee* against the grain, for Gissing extends little sympathy to Jessica Morgan *per se*. Much of the time she functions purely as an index of the worst aspects of Camberwellian pseudo-respectability. But in his next project, the novel that started out as *Benedict's Household* in late April 1895 and eventually saw light of day as *The Whirlpool*, Gissing takes hysteria and degeneration more seriously than he had ever allowed himself to do before. This may be a major reason why he found *The Whirlpool* so difficult to write, though, to be sure, the not wholly unrelated question of his marital difficulties also played a significant part. His *Diary* indicates that he found the initial labour of writing 'very hard' and, on the evening of 26 August, some four months after he had originally started work, he 'sat down and made a new beginning of my novel'. His new opening failed to resolve his difficulties, however, and even when he was just over thirty pages from the end Gissing complained of his debilitating 'slowness nowadays' and doubted whether he would 'ever get done'. Completing the novel on 18 December, he pronounced *The Whirlpool* '[t]he hardest bit of work I have done yet', notwithstanding his having written the bulk of the manuscript in slightly less than six months (Coustillas, pp. 409-429).

Gissing's abiding sense of Sisyphean toil, punctuated by moments of complete stasis, contrasts sharply with the novel's preoccupation with the uncontrollable force of the social maelstrom, modernity's fast-flowing rip-tide, at once deeply trivializing and utterly devastating – though both the process of composition and the published text produce, in their different ways, tales of setbacks and failures. But the perils of the 'whirlpool' are not unique to the 1897 novel of that name. They are also prefigured in a revealing comment made by Luckworth Crewe as he chaperons Nancy through the Jubilee crowds in Gissing's last triple-decker. Familiarly instructing her to take his arm as they prepare to move through Tottenham Court Road, Crewe warns Nancy: ' "These big crossings are like whirlpools; you might go round and round, and never get anywhere." ' As if to remind her of the lurking presence of violent elements, he adds: ' "Don't be afraid; if any one runs up against you, I'll knock him down" ' (*Jubilee*, p.70). Crewe's warning insinuates an image of the metropolis as harsh and futile, possessed of an empty or redundant dynamism, perpetual motion to no purpose. It is an image that resonates with the sordid struggle for existence in the book, as well as with the restless bustle of advertising, and suggests a world moved by a set of pre-social laws and conditions that individuals will never fully be able to steer or subdue. In embryo, this is the picture that *The Whirlpool* amplifies and exploits. Yet the range of meanings that it bears is extended and radically transformed as the word and its cognates become a constant refrain that operates like a metacommentary on the intersecting lives of the various characters.

Scattered somewhat artlessly throughout the novel, references to the 'whirlpool' of the title carry a number of overlapping significations. In the early chapters, the word is used to describe the fallout from a financial venture that has peaked and crashed in 1886, resulting in the suicide of the financier Bennet Frothingham and the ruin of those middle-class investors who had placed their trust in the 'wide net' offered by his expansively-named Britannia Loan, Assurance, Investment, and Banking Company, Limited (*Whirlpool*, p.12). Following the firm's collapse, Harvey Rolfe, the book's central male character, writes to his friend Hugh Carnaby that he feels as though 'we were all being swept into a ghastly whirlpool which roars over the bottomless pit', a nightmare vision of a kind of hellish vacuum lacking any final resting place that eerily echoes and generalizes Frothingham's suicide (p.48). In Rolfe's sympathetic note the 'whirlpool' intimates the dissolution of 'human fellowship', pointing towards an emptiness that lies on the far side of the drawing room and the gentleman's club, as if polite society were constantly in danger of being drained of all meaning. Rolfe has in fact already experienced a presentiment of this sudden reversal in the fortunes of the 'great finance-gambler' and his reactions anticipate the unravelling of the social world that will soon ensue. At one of Mrs. Frothingham's soirées the previous evening, he ponders the 'monstrous cruelties and mendacities' that 'might underlie the surface of this gay and melodious existence', dismissing its ritual niceties as 'the merest chatter' or the 'meaningless repetition of commonplaces' (pp. 39ff.).

Repetition – especially as embodied in the feverish cycle of socializing and conviviality that partially blinds its adherents to their follies and deceits – becomes a defining characteristic of what one of its victims calls 'the whirlpool way of life' (p.156). ' "Round and round they go," ' exclaims Rolfe, ' "brains humming till they melt or explode" ' (p.157). The allusion is overblown, simultaneously comic and sadistic. But this wild flight of fancy is addressed to Mary Abbott, who besides being a reader of Ribot has become the widow of one of Rolfe's London friends whose death from an overdose of morphia can be directly attributed to the Britannia Company crash. And part of the function of Rolfe's exultant words is to affirm the possibility of their standing outside the madcap flow. Looking back at the metropolis from the safe and timeless distance afforded by the Welsh countryside, these sometime Londoners see a world inhabited by 'the furiously busy' with 'no leisure, no meditation, no peace and quietness', a world that is urbane, self-absorbed and entirely meretricious, a world whose movement has changed both their lives forever (p.157). These references to a soulless, involuted social network, expanding and contracting according to the spasms of the financial system, but charged with its own distinctly entropic forms of energy, mark a return to Helmholtz's scientific *Weltanschauung* that is played out in several different registers. London society is at once a 'vortex of ... idle temptations' and a heady incitement (p.191). Thus Frothingham's daughter Alma, whose story forms the core of novel, can speak 'in her original, amusing way' of living 'on the outer edge of the whirlpool' (p.188). Here reputations wax and wane, sliding into an ostracism or social bankruptcy quite as destructive as the plunge in share prices when a speculative bubble bursts.

Alma's story – her subsequent marriage to Harvey Rolfe and her complex zigzagging between her family and her social or sexual pursuits – also intimates a whirlpool with a more personal but no less deadly rhythm, the fluctuations exhibited by the case history of a protracted bout of hysteria. In Alma R.'s case, hysteria ends in suicide and this climax effectively brings the novel full circle in a reprise of her father's suicide after the collapse of his financial empire. Alma is very much her father's daughter, the inheritor of his pathology, and it is implied that her distraction, her inability to settle properly into any given social role, shows a similar lack of will-power or moral fibre. But at the same time, her shallow ambitiousness, coupled with her lack of discipline or dedication, is of a piece with the high-toned social world in which she moves. There is in Alma's tragic career a perilous interplay between the disorders of the nervous body and the rather desperate superficiality of the *beau monde*. For example, because Alma's modest musical talent as a violinist is little more than a means to 'a place of distinction above ordinary handsome girls and heiresses', her 'emotional nature' becomes attuned 'to almost any kind of excitement' rather than to her art (p.246). And when, in a further twist of this vicious circle, her impulses drive her awry, she descends into 'a vortex of wretchedness' in which she has no memory of her actions, as though she had experienced an attack of fugue and had unknowingly wandered into strange territory (p.371). This dialectic of striving and loss reduces the broad pattern of her life to a species of neurasthenia. In incident after incident, Alma throws herself into a new enthusiasm, whether for drawing or mothering, for London sociability or rural seclusion, temporarily losing herself in an idealized activity whose attraction undergoes an inevitable attrition, leaving her depleted and hopelessly adrift. In a final 'crash', humiliated and directionless, she takes a fatal extra dose of 'her remedy for insomnia' in search of release from the 'great anguish of mind' that has come upon her – in a feminine version of Edgar Abbott's overdose following her father's suicide, fusing together the traces of these two despairing deaths (p.448).

As this brief outline suggests, the parallels between Alma's fate and the suicides occurring in the aftermath of her father's company's collapse point towards a close connection between financial depredations and the vicissitudes of the human body, perhaps implying that they are subject to a common logic. Certainly in the person of the millionaire Cyrus Redgrave, whose unintended death at the hands of Hugh Carnaby signals Alma's final nemesis, economic and sexual intrigue are tightly intertwined. Investment always comes freighted with ulterior motives. When Redgrave puts capital into Carnaby's bicycle manufacturing business it is not only to turn a profit, but also to compromise and so gain a hold over Carnaby's wife Sybil, who has secretly become his mistress. Similarly, when Alma is lured to Redgrave's Wimbledon bungalow on a rather dubious pretext, she discovers a place whose opulence 'at once allured and alarmed her', feelings that are underscored when his agent Mrs. Strangeways comments upon the expensiveness of each individual painting in the house. Mrs. Strangeways' remark 'Oh, he is immensely rich – and knows so well what to buy!' adds sexual innuendo to connotations of taste, implying that Redgrave can always buy whatever or whomever he wants (p.200). Significantly, the language of Alma's ambivalent response to the luxuriousness of Redgrave's bungalow in fact echoes that of an earlier episode in the book when, still smarting from the disgrace and misfortune that had befallen her family, she is invited to become his concubine. At first she is shocked, but then – though she cannot bring herself to accept – what was 'an affront' slips imperceptibly into 'an allurement' (p.87). Like Alma's father, Redgrave is one of the 'great speculators' and like Bennet Frothingham he too is addicted to living dangerously; but for Redgrave risk is always sexualized and, in courting sexual danger, he comes to an unexpected and ignominious end, killed by a jealous man he had thought to dupe (p.235). Before this happens, however, one might say that Redgrave succeeds in initiating Alma into the eroticism of risk, bringing out the bent or inclination that was already dormant within her nature.

Degeneration and manipulation are the twin keys to *The Whirlpool*, carried over from *In the Year of Jubilee* and substantially reworked. It is instructive to note that the up and coming young men in both books – Luckworth Crewe from *In the Year of Jubilee* and the popular composer Felix Dymes in *The Whirlpool* – are brash, unstoppable individuals without any pedigree. It is an essential part of their personae that they seem to come from nowhere, or nowhere very grand, and vulgarly elbow their way to attention and success. Representing rising commercial interests in leisure and advertising and well-versed in the exploitation of popular taste, the pure vigour of these unabashed and relatively unreflective men instantiates a new brutalism, a crude vitality that may lack discrimination or tact but which embodies a force that will drive forward the economic system and, *faute de mieux*, society as a whole. Part of the judgement against these men is that they each make a frank and altogether guileless play for the hand of the heroine and are each rebuffed. There is a social position, a social entitlement that is necessarily denied them, even though Crewe and Dymes really are the system's future. But the judgement is a complex one for, in resisting the lure of these men, the women in these texts are condemned to tread an awkward and damaging path. Nancy Lord is compelled to 'rationally' acquiesce to Tarrant's pseudo-Godwinian scheme of marriage in which the two lovers are bound to each other, yet pledged to live apart – a union which includes the stipulation that Nancy abandon any hope of a career as a popular novelist; while, for all Alma's attempted changes of direction, she is never able to escape the flawed constitution bequeathed by her father and must ultimately follow him into a vicious circle of recklessness, scandal, and suicide. Moreover, in *The Whirlpool* the question of social regeneration is deliberately posed in the bleakest possible terms and can neither be settled nor exhausted by Alma's death alone. References to the 'brute savagery' of Kipling's *Barrack-Room Ballads* in the final chapter, jarringly described by Harvey Rolfe as ' "The tongue of Whitechapel blaring lust of life in the track of English guns!" ', do nothing to suggest a way forward. For although at the time these remarks were taken by some critics to be an endorsement of the expansionist imperial project on Gissing's part, Rolfe's sentiments are in fact equivocal rather than wholeheartedly martial (Gettmann, pp. 48f.; and Simon J. James in the present volume notes Gissing's disapproval of Kipling's influence). What makes Rolfe's unease especially poignant, however, is the lack of any consolation from within his own family. Insofar as Rolfe's son Hughie holds the future in his hands, this 'slight', 'wistful', 'timid', 'weary' child seems to conjure scant promise of solid national advancement (pp.449ff.). Between the bullying jingoistic *ressentiment* of the music hall and the listless child who still needs to 'learn to be noisy' there seems to be no viable alternative (p.452). Revealingly, Hughie has been named after Hugh Carnaby, the spirited imperial settler who becomes a kind of male hysteric, 'tottering like a feeble old man' under the unforgiving gaze and command of his ruthless wife (p.439).

Freud and Gissing: the diagnosis of 'nervous disease'

One of the strengths of Freud's work is that he took as his starting-point a refusal of the degenerationist argument that bulked so large in Gissing's oeuvre and which reached a climacteric in *The Whirlpool*. Freud's case against his great teacher Charcot was that he so overestimated 'heredity as a cause that no loophole was left by which nervous disease could be acquired' and that in consequence he was unable to 'differentiate sharply enough between the organic nervous affections and the neuroses' (Freud, 1963, pp.24f.). Curiously enough, exactly the same criticism can be levelled at Gissing's later novels. On the one hand, characters like Nancy struggle to escape their patrimony and, in appropriately guarded terms, partially succeed, though their degenerative family background dogs their every step; while on the other, Alma represents the degenerate condition at its most hopeless, in which every encounter only serves to reinforce or release her worst traits, and every apparent choice produces a new fatalism. In the peaks and troughs of Alma's psychic life, in her turbulent hopes and bitter anti-climaxes, one begins to see a relentless unbinding that resembles neurasthenic breakdown, or worse still a nature working its way through to its own extinction, as predicted by the second law of thermodynamics. And from another perspective, one could also take Alma's life story to be a manifestation of the death drive.

In a sense, these bleak dichotomies might be said to be homologous with certain features of Gissing's own biography, features that may be articulated into a personal myth: the desire to rescue or rehabilitate a woman without qualities, coupled with a deep instinctive pessimism as to the very possibility of such a gesture. But if one takes Freud's theoretical innovations as a benchmark, the instabilities in Gissing's later novels gradually become clearer. As has already been noted, psychoanalysis took as its Archimedean principle a differentiation between those types of mental disturbance that were tied directly to somatic processes, and those that displayed 'a complicated psychical mechanism'. Among the former were neurasthenia, which Freud saw as the product of sexual excess, and what he called 'anxiety neurosis' whose '*contrary* aetiology' stemmed from 'an accumulation of physical [i.e. sexual] tension' (Breuer and Freud, p.340). The psychoneuroses, by contrast, included hysteria and obsessional neurosis in which the appearance of physical symptoms could be traced back to mental conflicts without any corresponding link to bodily lesions or physiological dysfunctions. Freud's classification marked a new beginning in therapy and in theory, for out of it emerged several of psychoanalysis's most distinctive concepts: repression, fantasy, and the unconscious. Their aggregate effect was to slacken the knot commonly thought to bind the physiology of desire to its objects: in short, to deny that there are innate predispositions that naturally determine the relation between human instincts and their satisfactions.

One aspect of the radical contingency of sexual desire upon which Freud placed particular emphasis was its malleability or versatility, the ability to change its aim without losing any of its 'intensity'. Precisely because he believed that the sex drive was 'probably more highly developed in human beings than in most of the higher animals' or at least 'more constant', Freud thought that it was the source of immense reserves of energy whose successful management and deployment constituted civilized society's greatest asset and also its greatest problem (Freud, 2002, p.90). In his 1908 essay on ' "Civilized" Sexual Morality and Modern Nervous Illness' Freud argued that Western civilization had largely been achieved through curbing and re-directing the various components of the sex drive, laying down standards of acceptable social behaviour and tacitly encouraging non-sexual outlets in lieu of a potentially unbridled eroticism. However, in contrast to critics like Krafft-Ebing or Erb who claimed that the modern rise in nervous diseases was primarily due to the unremitting pace of commercial and industrial activity, reaching a fever pitch in the widespread political and economic crises of the day, Freud insisted that the root cause of this phenomenon could be traced to the difficulties stemming from the many taboos against the pursuit of immediate sexual gratification. The ductility of the sex drive was by no means boundless and many attempts at taming or transforming it were counter-productive, resulting in 'an increase in our anxiety about life and our fear of death' and a decline in the spirit of patriotic self-sacrifice (p.104). Far from possessing infinite plasticity, Freud believed that there were strict limits on the extent to which sexual desire was amenable to sublimation or displacement, 'just as' – to use a very Helmholtzian analogy – 'there is a limit to the conversion of heat into mechanical energy in our machines' (p.91). In Freud's view – and this was the main lesson to be drawn from his early clinical practice – the costs incurred through 'the suppression of our drives' were now clearly outstripping the possible benefits (p.90).

Like Freud's case histories, Gissing's later novels oscillate between tension and excess, between desire and repression. But, while the latter's major characters also show signs of 'a complicated psychical mechanism' governing their everyday lives, Gissing often seems unsure as to how much responsibility they can bear for their actions. Alma, for instance, represents a remarkably complicated case of someone who is as thoroughly bound by her 'hereditary shame' as it is possible to be (*Whirlpool*, p.419). When the threat posed by an enemy releases in her 'an instinct of savagery' – 'in a flash of primitive mind, she saw herself spring upon her enemy, tear, bite, destroy' – far from acting upon this murderous impulse, she walks the streets 'like a somnambulist', completely shaken by her archaic yet unrealizable desire (p.428). Her tragedy is constituted not merely by the fact that she is the daughter of a man 'who had played for the world's reward, and, losing, flung away his life' and that she is barred forever from 'the pleasures of a dull, decorous circle', but that she *knows* that this is her fate (p.419). In this sense, her inner life shows her to be more of a sister to Nancy Lord than she might at first seem. Yet if, as Gissing evidently believed, '[t]he law of Evolution of course affects heredity' so that '[t]he individual ceaselessly tends to modify himself', Alma's case also reveals the limits of its jurisdiction, notwithstanding her brief moments of self-control (Korg, p.61). While she fitfully endeavours to check or transform the 'troublesome mutability of temper' (p.403) that indicates her basic lack of discipline and purpose, even striving to 'hypnotize her imagination with a new ideal of life' in desperate, unwitting simulation of psychoanalysis's opening therapeutic moves (p.423), her efforts are to no avail. In her dreams, as in her waking life, Alma is plagued by a past that she can never truly overcome: before she falls asleep she imagines that she can hear the dead Cyrus Redgrave's voice, faithfully reproducing 'his tricks of tone, his suavities of expression', and once asleep she sees him looking at her with 'a smile of such unutterable malignity that she froze with terror' (p.369). A helpless sufferer from reminiscences, Alma's involuntary spasms mime an attempted escape from self-destruction as much as they violently prefigure her own end: the 'loud hysterical cry; then sobs and wailing, with movements as if to tear open the clothing that choked her' (p.443).

Gissing's adherence to a degenerationist view of Britain's decline could be harsh and uncompromising. 'One cause of national degradation, now-a-days, is the success of medicine in keeping alive the unfit', he once wrote (Korg, p.61). All the more reason therefore that where the difficulties and contradictions of such a view start to surface in his novels he was singularly ill-equipped or unwilling to deal with their implications. But Gissing's inability to address these problems was partially obscured by the presence of a very different schema in his work, one that allowed him to put his finger on a form of pathology that Freud neglected or – thinking here of the hardship Freud experienced as a result of the failure of his own father's business – perhaps repressed. While Freud's extensive use of economic metaphors, such as *Gewinn* (yield, profit) or *Besetzung* (investment), pictured the self as a kind of capitalist machine, his interest in the fate of the individual *within* the wider economy was noticeably underdeveloped. Gissing was equally preoccupied with individual lapses and limitations, but at the same time his depiction of a whole society in 'crisis' – a term that he applied indiscriminately to the personal and the economic alike – illuminates a world in which suffering is systemic, in which people's worst habits and meanest ambitions are not simply stifled by conventional morality, but rather are channelled into and exacerbated by the workings of an institutional order whose inducements and malfunctions they are subsequently unable to countermand. This is a world that is represented by such ventures as the Britannia Loan, Assurance, Investment, and Banking Company on the one hand and by the raw new stage of imperialism that began in the 1880s on the other. Neurasthenia and hysteria may become hopelessly confused in Gissing's work and his haphazard hereditarianism may lie some way behind Freud's more rigorous insights. Yet in the final analysis the whirlpool provides a metaphor not only for the helpless rush of depressive feelings in the face of disaster, but for the terrifying experience of being caught up in a network of implacable and unrelenting material forces, forces with a reach that has grown ever more global in scope. For this condition there can be no talking cure. But whether the terms of Gissing's own grim diagnosis can enable any humane practical remedy remains, alas, a far more troubling issue.

Bibliography

Bowlby, Rachel (1985), *Just Looking: Consumer Culture in Dreiser, Gissing and Zola*, Methuen, New York and London.

Breuer, Josef and Sigmund Freud (1974 [1895]), *Studies on Hysteria*, Pelican Books, Harmondsworth.

Coustillas, Pierre (ed) (1978), *London and the Life of Literature in Late Victorian England: The Diary of George Gissing, Novelist*, Bucknell University Press, Lewisburg.

Freud, Sigmund, ed. Philip Rieff (1963), *Early Psychoanalytic Writings*, Collier Books, New York.

Freud, Sigmund (2002), *Civilization and Its Discontents*, Penguin Books, Harmondsworth.

Gettmann, Royal A. (ed.) (1961), *George Gissing and H.G. Wells: Their Friendship and Correspondence*, University of Illinois Press, Urbana.

Gissing, George (1980 [1895]), *Eve's Ransom*, Dover Publications, Toronto and London.

Gissing, George (1987 [1894]), *In the Year of Jubilee*, Hogarth Press, London.

Gissing, George, ed. Arlene Young (1998; first publ. 1893), *The Odd Women*, Broadview Press, Peterborough, Ontario.

Gissing, George (1984 [1897]), *The Whirlpool*, Hogarth Press, London.

Goode, John (1978), *George Gissing: Ideology and Fiction*, London, Vision Press.

Greenslade, William (1994). *Degeneration, Culture and the Novel*, Cambridge University Press, Cambridge.

Helmholtz, Hermann von (1873-1881), *Popular Lectures on Scientific Subjects*, 2 series, Longmans, Green and Co., London.

Korg, Jacob (ed.) (1962), *George Gissing's Commonplace Book*, The New York Public Library, New York.

Oppenheim, Janet (1991), *"Shattered Nerves": Doctors, Patients, and Depression in Victorian England*, Oxford University Press, New York and Oxford.

Pick, Daniel (1989), *Faces of Degeneration: A European Disorder, c.1848-c.1918*, Cambridge University Press, Cambridge.

Rabinbach, Anson (1992), *The Human Motor: Energy, Fatigue, and the Origins of Modernity*, University of California Press, Berkeley and Los Angeles.

Stedman Jones, Gareth (1971), *Outcast London: A Study in the Relationship between Classes in Victorian Society*, Clarendon Press, Oxford.

Sulloway, Frank J. (1992), *Freud, Biologist of the Mind: Beyond the Psychoanalytic Legend*, Harvard University Press, Cambridge (Massachusetts) and London.

Chapter 6

The Discontents of Everyday Life: Civilization and the Pathology of Masculinity in *The Whirlpool*

Simon J. James

If you can keep your head when all about you
Are losing theirs and blaming it on you
If you can trust yourself when all men doubt you
But make allowance for their doubting too …

If you can fill the unforgiving minute
With sixty seconds' worth of distance run,
Yours is the Earth and everything that's in it,
And – which is more – you'll be a Man, my son!
Rudyard Kipling, opening and closing lines of 'If'

'When one has children what one writes becomes a terrible responsibility.'
Miss Ambient, in Henry James, 'The Author of Beltraffio'

Gissing and masculinity

In his fiction of the 1890s, Gissing's attention becomes increasingly engaged by issues of Empire and national identity, and, to a greater extent than previously, by late Victorian culture's construction of gender. Gissing had been preoccupied to some extent with the nature of masculinity since his first novel *Workers in the Dawn* (1880). The hero Arthur Golding tries and fails, like Henry James's effeminately-named Hyacinth Robinson (in *The Princess Casamassima*), to square external demands for masculine action with an internalized responsibility for the production of art. Gissing continued to chart the ways in which the male protagonist's identity might be threatened or oppressed, most often by the balance of power within marriage (Grylls 1986, pp.162f.). In *New Grub Street* (1891), Edwin Reardon struggles to assert himself against his 'masculine' wife:

> He had but to do one thing: to seize her by the arm, drag her up from the chair, dash her back again with all his force – there, the transformation would be complete, they would stand towards each other on the natural footing. With an added curse perhaps – Instead of that, he choked, struggled for breath, and shed tears.
>
> Amy turned scornfully away from him. Blows and a curse would have overawed her, at all events for the moment; she would have felt: "Yes, he is a man, and I have put my destiny into his hands." His tears moved her to a feeling cruelly exultant; they were the sign of her superiority. It was she who should have wept, and never in her life had she been further from such display of weakness. (*New Grub Street*, pp.261f.)

There is an analogous passage in *Isabel Clarendon* (vol.2, p.217.) In *The Whirlpool* (1897), and also the later anti-imperialist novel *The Crown of Life* (1899), Gissing concentrates on the location of a masculinity precariously defined between an internally-valorized conception of self and external pressures towards politics, violence, displayed Englishness and manly activity (Mangan and Walvin). In *The Whirlpool* especially, masculinity is threatened by urban consumerist pressures that Gissing genders as female; the Empire provides a possible outlet for sublimated masculine urges, but not without cost to the development of the higher self. In 'Women and the Disease of Civilisation', William Greenslade has identified *The Whirlpool*'s project to establish a 'nervous pathology' of modern living (Greenslade, pp.134-150), a theme further discussed by David Glover in the present volume. The novel's heroine, Alma Frothingham, who becomes Alma Rolfe, is sucked into the 'whirlpool way of life' in and around London (*Whirlpool*, p.156). She is morally compromised by her desire for both a commodified, public modernity in her career as a musician and a traditional family existence as a wife. 'The delight of an existence that loses itself in whirl and glare' has its cost, however, and Alma eventually dies from an overdose of a remedy for 'fashionable disorder of the nerves' (p.305). As H.G. Wells, among others, has noted, Gissing tends to group the characters of his novels around a particular 'structural theme': in this novel the theme is 'the fatal excitement and extravagance of the social life of London' ('The Novels of Mr George Gissing', in Gettmann, pp.242-259; see p.245). The male characters who risk being propelled into the whirl of modern living by their hysteric womenfolk are thus also made subject to the same nervous disorders as the women – although the women's symptoms tend to be more conspicuous (see Showalter; and Federico, pp.106f. My reading of Rolfe's development during the course of the novel, however, differs from Federico's). After he has been dealing in the City, 'beneath [Hugh]'s eyes hung baggy flesh that gave him a bilious aspect, his cheeks were a little sunken, and the tone of his complexion had lost its healthy clearness' (*Whirlpool*, p.209); Abbott suffers from neuralgia (p.7); Morphew from 'indifferent health' (p.90); Rolfe from a sore throat (p.209); his father from fever (p.330).

Gissing is still wrongly portrayed as an unthinking supporter of patriarchy (for a recent account of this issue, see Kramer). Although certainly capable of misogynist sentiments, in novels such as the now better-known *The Odd Women* (1893) Gissing can demonstrate an awareness of how the oppression of women is less determined essentially by their sex than culturally by their gender, usually through unsatisfactory education. This is the argument of an often-quoted letter he wrote in 1893:

> My demand for female 'equality' simply means that I am convinced that there shall be no social peace until women are intellectually trained very much as men are. More than half the misery of life is due to the ignorance & childishness of women. The average woman pretty closely resembles, in all intellectual considerations, the average male *idiot* – I speak medically. That state of things is traceable to the lack of education, in all senses of the word. Among our English emancipated there is a majority of admirable persons.... I am driven frantic by the crass imbecility of the typical woman. That type must disappear, – or at all events become altogether subordinate. And I believe that the only way of effecting this is to go through a period of what many people will call sexual anarchy. (Mattheisen et al, vol.5, p.113)

While, of course, women are portrayed in Gissing's work as more oppressed by the limitations of their gender than men, Gissing could perceive the constructed nature of masculinity as well as femininity. Often in fiction of the *fin-de-siècle* by male authors such as Stevenson, Wilde, Rider Haggard, Conan Doyle and Du Maurier, literary presentation of masculinity tends to become a field of enquiry rather than a single node, offering versions within an identifiable typology, especially through contrasting pairings. *The Whirlpool* also offers this kind of sexual typology. Set in contrast to the central male protagonist Harvey Rolfe are his friends Cecil Morphew, Hugh Carnaby, and Basil Morton, who display different modern – or in the case of the last, archaic – 'types', and who before Rolfe's marriage comprise a network of homosocial relationships, 'quaffing and smoking and talking into the small hours' (*Whirlpool*, p.128). Carnaby, for instance, is recognizably of the stock which peopled in life Victoria's, and in literature, Kipling's, Empire (Gatrell, pp.79f):

> Carnaby was a fair example of the well-bred, well-fed Englishman – tall, brawny, limber, not uncomely, with a red neck, a powerful jaw, and a keen eye. Something more of repose, of self-possession, and a slightly more intellectual brow, would have made him the best type of conquering, civilising Briton. He came of good family, but had small inheritance; his tongue told of age-long domination; his physique and carriage showed the horseman, the game-stalker, the nomad. Hugh had never bent over books since the day when he declined the university and got leave to join Colonel Bosworth's exploring party in the Caucasus. After a boyhood of straitened circumstances, he profited by a skilful stewardship which allowed him to hope for some seven hundred a year; his elder brother, Miles, a fine fellow, who went into the army, pinching himself to benefit Hugh and their sister Ruth. Miles was now Major Carnaby, active on the North-West Frontier. Ruth was wife of a missionary in some land of swamps; doomed by climate, but of spirit indomitable. It seemed strange that Hugh, at five and thirty, had done nothing particular. Perhaps his income explained it – too small for traditional purposes, just large enough to foster indolence. For Hugh had not even followed up his promise of becoming an explorer; he had merely rambled, mostly in pursuit of fowl or quadruped. When he married, all hope for him was at an end. The beautiful and brilliant daughter of a fashionable widow, her income a trifle more than Carnaby's own; devoted to the life of cities, wherein she shone; an enchantress whose spell would not easily be broken, before whom her husband bowed in delighted subservience – such a woman might flatter Hugh's pride,

> but could scarce be expected to draw out his latent energies and capabilities. This year, for the first time, he had visited no wild country; his journeying led only to Paris, to Vienna. In due season he shot his fifty brace on somebody's grouse-moor, but the sport did not exhilarate him.
>
> An odd and improbable alliance, that between Hugh Carnaby and Harvey Rolfe. Yet in several ways they suited each other.... The one's physical vigour and adroitness, the other's active mind, liberal thoughts, studious habits, proved reciprocally attractive. Though in unlike ways, both were impressively modern. Of late it had seemed as if the man of open air, checked in his natural courses, thrown back upon his meditations, turned to the student, with hope of guidance in new paths, of counsel amid unfamiliar obstacles. (*Whirlpool*, pp.8ff.)

Gissing initially polarizes hero and hero's best friend between, in a phrase that Joseph Bristow has used writing on E.M. Forster, 'the opposed interests of the sociable sportsman and the smug intellectual' (Bristow, p.61). Carnaby is only partly ironized here (unlike the violent imperialists the Strattons in *Isabel Clarendon*). While recognizing Carnaby's character to be of a different kind from his own, Rolfe can potentially see a place in the world for the vigorous, unscholarly sportsman. Such a place might ideally, as for Carnaby's siblings, be the Empire: instead he is unnaturally confined to the enclosed spaces of London society by his marriage to a woman with a 'decided preference for civilisation' (p.34; on the tension between marriage and manly imperial activity, see Tosh, p.176).

A man's place

This novel's anxieties are confined mostly within the domestic sphere. While Gissing evidently feels that the home should largely be the concern of women, this household should still be ruled by masculine power, a view expressed in the idealized Mrs Morton's commonplace that ' "A married woman would, of course, be guided by her husband's wish" ' (*Whirlpool*, p.337; cf. Mattheisen et al, vol.6, pp.149f.). The question most urgently debated by the novel's married couples is where this home should be. The novel dramatizes a choice between a static, rural ideal supported by individual female service, or continuous and debilitating urban excitement driven by collective feminine desire (see James, 1997, pp.129-138). The Empire supplies the possibility of a space further beyond Europe where the potentialities of the masculine self could be given fuller expression. Unlike many nineteenth-century representations of overseas territories, Gissing's portrayal of the Empire does not feminize it as fertile foreign territory awaiting colonization and domestication by a vigorous English male, but rather presents it as a liminal space where manly energies can be exerted with the least damage to metropolitan, feminine English society.

> "Why didn't I follow Miles into the army? I think I was more cut out for that than for anything else. I often feel I should like to go to South Africa and get up a little war of my own."
>
> Rolfe shouted with laughter.
>
> "Not half a bad idea, and the easiest thing in the world, no doubt."

"Nigger-hunting; a superior big game."

"There's more than that to do in South Africa," said Harvey. "I was looking at a map in Stanford's window the other day, and it amused me. Who believes for a moment that England will remain satisfied with bits here and there? We have to swallow the whole, of course. We shall go on fighting and annexing, until – until the decline and fall of the British Empire. That hasn't begun yet. Some of us are so over-civilised that it makes a reaction of wholesome barbarism in the rest. We shall fight like blazes in the twentieth century. It's the only thing that keeps Englishmen sound; commercialism is their curse. Happily, no sooner do they get fat than they kick, and somebody's shin suffers; then they fight off the excessive flesh. War is England's Banting."

"You'd better not talk like that to Sibyl." (*Whirlpool*, pp.16f.)

Before his marriage, Rolfe happily travels on his own in Europe, this foreign space one of contented masculine leisure; subsequently his freedom to travel is restricted by marriage and parenthood. Mrs Morton, apparently with the narrator's approval, is happy to remain confined to England unless safely escorted by male relatives: ' "I will go abroad," she said, "when my boys can take me" ' (p.325). Before the correction of his initial views by the actual fact of fatherhood, Rolfe mistakenly speculates:

"If I had a son," pursued Harvey, smiling at the hypothesis, "I think I'd make a fighting man of him, or try to. At all events, he should go out somewhere, and beat the big British drum, one way or another. I believe it's our only hope. We're rotting at home – some of us sunk in barbarism, some coddling themselves in over-refinement. What's the use of preaching peace and civilisation, when we know that England's just beginning her big fight – the fight that will put all history into the shade! We have to lead the world; it's our destiny; and we must do it by breaking heads. That's the nature of the human animal, and will be for ages to come."

Carnaby nodded assent. (p.14)

As *The Crown of Life* later shows, Gissing can certainly be critical of the unappealing cultural consequences of imperialism, but for certain types of men, the Empire might yet be seen as the most suitable place for them, where they may temporarily 'decivilise' themselves, work off their 'superabundant vitality as nature prompted' (p.45) and exert masculine urges that this novel's discourse constructs as 'natural instincts' (p.17). Nature and civilization are thus placed by this novel in opposition, in particular masculine 'nature' and feminine, urban 'civilization'.

Addressing these concerns in the aftermath of the First World War, Sigmund Freud wrote:

The inclination to aggression is an original, self-subsisting instinctual disposition in man, and I return to my view that it constitutes the greatest impediment to civilisation.... I may now add that civilisation is a process in the service of Eros, whose purpose is to combine single human individuals, and after that, families, then races, peoples and nations, into one great unity.... But man's natural aggressive instinct, the hostility of each against all and of all against each, opposes this programme of civilisation. (Freud, p.122)

Freud blames the unsuccessful sublimation of this violent masculine urge for the wars which threaten the continued survival of nations and the species, and credits 'higher psychical activities, scientific, artistic or ideological' to successful sublimation (p.97). Rolfe is too temperamentally studious to need a violent outlet, but his unlettered friend Carnaby does. The narrator repeatedly indicates that the frustrated Empire-builder is unnaturally confined by his marriage, and by the urban living it demands; he contracts 'an odd habit of swinging his right arm, with fist clenched, as though relieving his muscles after some unusual constraint' (*Whirlpool*, p.276).

Love and marriage

Male identity, for Gissing, is not given expression by the sexual act whether in or outside of marriage: masculinity is conquered and restrained by sex. When Rolfe first falls for Alma, 'all his manhood [is] subdued by her scornful witchery' (this phrase occurs in a passage of very poor writing, as if Gissing's own capabilities are subdued by the difficulties of representing a powerful female sexuality). Carnaby is still more baffled, and by other women besides his wife (*Whirlpool*, p. 102). He is robbed by his housekeeper, Mrs Lant, of his symbolically masculine possessions of silver-mounted pistols, cheque-book and ibex-hoof paperweight. After losing money in the Britannia crash, he goes abroad, temporarily to the open space of Queensland but then to an émigré version of the civilized London society he would prefer to escape, in Honolulu.

> The tropics were not his favourite region, and those islands of the Pacific offered no scope for profitable energy; he did not want to climb volcanoes, still less to lounge beneath bananas and breadfruit-trees, however pleasant such an escape from civilisation might seem at the first glance. A year of marriage, of idleness amid amusements, luxuries, extravagances, for which he had no taste, was bearing its natural result in masculine restiveness. His robust physique and temper, essentially combative, demanded liberty under conditions of rude or violent life. (p.58)

On their return, Sibyl's continued consumption having prevented any money being saved, Carnaby is obliged to enter a bicycle-manufacturing business. This allows some scope for his natural instincts: 'He preferred to stay down at Coventry with his partner Mackintosh, living roughly, smoking his pipe and drinking his whisky in the company of men who had at least a savour of sturdy manhood' (p.184). However, the novel seems to suggest that since the activity of trying to make money is one enforced by female consumption, it is inherently feminizing, or at least threatening to masculinity's (sexual) capital: 'his days of sport were gone by; he was risking the remnant of his capital' (p.184). Carnaby is thus obliged to seek additional 'capital' from the financier Redgrave. When by chance Carnaby re-encounters his former housekeeper, she hints that Sibyl has secured the loan by sleeping with Redgrave (the narrative withholds confirmation of whether this is true). Carnaby goes to Redgrave's suburban villa and, mistaking Alma's presence there for Sibyl's, strikes the millionaire, killing him.

Carnaby has stated earlier that 'men must not quarrel; women did more than enough of that' (p.279); but the physical alternative to verbal quarrelling proves more harmful still. In this novel at least, action is natural for men, but women are more at home in language, especially untruthful language. ' "I hate a dirty, lying, incapable creature, that's all, whether man or woman. No doubt they're more common in petticoats," ' claims Rolfe (p.15). Carnaby's forceful masculinity is ill-equipped to deal with the unreliable female discourses of gossip and slander:

> Hugh Carnaby – the man who had lived as high-blooded men do live, who had laughed by the camp-fire or in the club smoking-room at many a Rabelaisian story and capped it with another, who hated mock modesty, was all for honest openness between man and woman – stood in guilty embarrassment before his own wife's face of innocence. (p. 300)

The trip abroad and the time he spends in prison both make Carnaby less of a man – literally, as his physique wastes away, and socially when he becomes infantilized by his wife, who addresses him as her 'boy'. The novel seems to blame this episode on the nature of modern living: if society allowed men such as Carnaby the scope to exert 'natural' urges towards violence, such incidents would not occur; instead, Carnaby and Redgrave are victims of civilization's failure to accommodate their instincts of sexual competition.

> Men are not gentle creatures who want to be loved, and who at the most can defend themselves if they are attacked; they are, on the contrary, creatures among whose instinctual endowments is to be reckoned a powerful share of aggressiveness. As a result, their neighbour is for them not only a potential helper or sexual object, but also someone who tempts them to satisfy their aggressiveness on him, to exploit his capacity for work without compensation, to use him sexually without his consent, to seize his possessions, to humiliate him, to cause him pain, to torture and to kill him. (Freud, p.211)

'Civilization', in *The Whirlpool*'s fashionable London society, is less the culture which Rolfe is seen enjoying in pleasurable masculine seclusion at the novel's opening than an unnatural, feminizing, societal restraint. Although proximity to London allows Harvey temporary liminal escape to the world of the club – where he can turn, as Alma puts it, ' "into a sort of bachelor again... quite in a boyish way" ' (p.172) – the range of its amusements is insufficient to satisfy the outdoor-minded Carnaby, making him 'get quarrelsome and wish he was on the other side of the world' (p.213). He complains, in a flat over-crowded by feminine possessions:

> "What *should* I do? I'm not the man for books; I can't get much sport nowadays: I don't even care for billiards. I want to have an axe in my hand!"
>
> Gesticulating carelessly, he swept a wine-glass off the table. (p.215)

A different kind of 'accident' resulting from a 'natural' masculine urge blights the life of another of Rolfe's friends, Cecil Morphew. Morphew's marriage to Henrietta Winter has been delayed for ten years because of the objections of his fiancée's parents to his straitened means. The death of her father makes the marriage possible, but Henrietta calls it off after learning of Morphew's liaison with a working-class girl, resulting in a child, now dead. Had it not been for lack of money (the constant thematic preoccupation of Gissing's fiction) and the Winters' concern with social appearances, Morphew's masculine sexual desires would have been directed sooner and correctly towards the woman he really loves (pp.348f.). Like Carnaby's urge towards robust violence, when checked, these energies find an outlet elsewhere. The shock of rejection nearly drives Morphew, the most nervous and hyper-sensitive of the male characters, towards the eternal sleep hinted at by his name. ' "There is a point in the life of every man with brains," ' (he declares earlier) ' "when it becomes a possibility that he may kill himself" ' (p.326). It is the sickly Henrietta, however, who prematurely dies instead, and rather than directing these urges harmfully inwards, Morphew is free to emigrate to the more hopefully manly territory of New Zealand.

Rolfe is contrasted with these and the novel's other men: Frothingham, the Merdle/Melmotte-like speculator who, true to literary type, commits suicide shortly before his exposure; Wager, who deserts his children when he loses his money in the Britannia crash; Leach, who risks death by labouring to fund his wife's and daughters' spending; Abbott, who *is* worked to death by his family; the sexually speculative Cyrus Redgrave; the successful vulgar composer Felix Dymes, who eventually marries an eighteen-year-old heiress. Rolfe attempts a marriage of more equal minds with Alma; however, not only the narrator but Alma herself hint that Rolfe may do well to exert his (literally) patriarchal authority upon her more than he does: 'Once or twice she had thought (perhaps had hoped) that he would lay down the law in masculine fashion; but no' (p.254; and see also pp.163, 166, 305). The novel never indicates that Rolfe's unspoken words should be followed by the proverbial blow. However, Sibyl does suggest that to be entrapped like him is ' "the fate of men – except those who have the courage to beat their wives" '. ' "Rolfe isn't half the man he was," ' agrees her husband (p.277). John Tosh has argued on the place of patriarchy in the nineteenth century:

> "Patriarchy" has become an unfashionable term in recent years, as indicating a crudely reductionist view of sexual stratification. But, in its precise meaning of 'father rule', patriarchy remains an indispensable concept, not only because men have usually wielded authority within the home, but also because it has been necessary to their masculine self-respect that they do so. (Tosh, p.3).

Parenthood: becoming a man

Rolfe ponders what it means to be a man not only because of his own uneasiness with 'masculine decision, ... the old-fashioned authority of husbands' (*Whirlpool*, p.382), but also because of his new role as a father, responsible for the type of man little Hugh Rolfe will grow into. Wells, in his review of *The Whirlpool*, noted the emergence of child-rearing as a new theme in Gissing's fiction (Gettmann, p.244). Gissing himself had become a father in 1891. He wrote in his diary for December 10th:

> 4.15 a.m. Have been up all night. A furious gale blowing. E[dith, his wife] in long miserable pain; the doctor has just given her chloroform, and says that the blackguard business draws to an end.
>
> 5.15. Went to the study door, and heard the cry of the child. Nurse, speedily coming down, tells me it is a boy. Wind howling savagely. So, the poor girl's misery is over, and she has what she earnestly desired. (Gissing, ed. Coustillas, 1978, p.265)

Although initially Gissing's reactions to his offspring shifted between indifference and pity, the welfare of his children soon became a constant presence in his diary and letters. (See, for instance, Gissing, ed. Coustillas, 1978, pp.296, 317, 326, 345, 396, 407, 441; and Mattheisen et al, vol.4, throughout). Gissing fretted constantly about his and Edith's capacities as parents, especially as their unsuitability as husband and wife became more apparent. In 1896, the year in which Alfred, his second son, was born and in which *The Whirlpool* was written, Gissing confided in his diary:

> [Walter] knows there is no harmony between his mother and me, and he begins to play upon the situation – carrying tales from one to the other, etc. The poor child is ill-tempered, untruthful, precociously insolent, surprisingly selfish. I can see that Wakefield [where Gissing's mother and sisters lived] *may* have a good influence, but only the merest beginnings show as yet. I should like to know how the really wise and strong father would act in this position. But no wise and strong man could have got into it. Talk of morals! What a terrible lesson is the existence of this child, born of a loveless and utterly unsuitable marriage. (Gissing, ed. Coustillas, 1978, p.418)

Gissing wrote to Wells: 'In Rolfe I wished to present a man whose character developed with unusual slowness, & who would probably never have developed at all, after a certain stage, but for the change wrought in his views & sentiments by the fact of his becoming a father' (Mattheisen et al, vol.6, p.320). Rolfe reflects self-consciously on this 'revolution in his thoughts' at the beginning of the novel's second part (*Whirlpool*, p.135). The reader is informed of all the circumstances which stunted Rolfe's development and from which he wants to save his own son, especially given his concerns about 'nervous tendencies inherited from his mother' and her mother before her (pp.330ff., 383; see also p.228). Much of Rolfe's identity becomes invested in his role as a father: 'fruitless his life could not be, if his child grew up' (p.139). In a letter to Thomas Hardy in 1895, Gissing wrote:

> One theme I have in mind – if I can ever get again to a solid book – which I want to treat very seriously. It is the question of a parent's responsibility. This has been forced upon me by the fact that I myself have a little boy, growing out of his infancy; the thought of his education – in the widest sense of the word – troubles me day & night. (Mattheisen et al., vol.6, p.20)

Gissing later wrote to Eduard Bertz that the theme of his new novel, initially called *Benedict's Household*, was 'the decay of domestic life among certain classes of people, & much stress is laid upon the question of *children*' (Mattheisen et al., vol. 6, p.123; see also the letter to H.H. Sturmer, pp.178f.). The 'matter of education' (*Whirlpool*, p.391) preoccupies the hero, as it did the author during Gissing's several changes of mode of living in the early years of his sons' lives. Even by the end of the novel, Rolfe has not entirely dismissed the suggestion that he should try to raise the bookish and weakly Hugh, possessed of 'nervous tendencies which were to be expected in a child of such parentage' (p.383), to be more like his namesake Hugh Carnaby, thus standing a better chance of survival in a world increasingly violent and jingoistic (there are analogous moments in *Isabel Clarendon* and *Born in Exile*: see pp.237ff. and p.458 respectively). Rolfe has no such doubts about the other boy for whose education he is responsible: Albert Wager, who 'loathes everything but games and fighting' (p.390), is sent into the navy.

Finally a lone male parent to his son, in this novel full of unsatisfactory mothers, Rolfe ultimately moves to Greystone, near Morton's idyllic household. (Significantly, the female children born to Alma and Mrs Abbott both die.) In the book's closing scene, Rolfe and Morton discuss Kipling's *Barrack-Room Ballads*. Many reviewers, including H.G. Wells, misunderstood Rolfe's apparent enjoyment of the 'strong man made articulate' (p.449). Gissing corrected Wells's review in a letter: 'I never meant to suggest that Rolfe tended to the "Barrack-Room" view of life. In all he says, he is simply expressing his hopeless recognition of facts which fill him with disgust' (Mattheisen et al., vol.6, p.320; for Gissing's own disapproval of the influence of Kipling, see vol.7, p.236). Such resigned 'recognition' is one of the strengths of Gissing's fiction, a successful deployment of the realist mode which allows the confrontation of material phenomena that the writer finds inimical, while (with the exception of *The Private Papers of Henry Ryecroft*) avoiding a utopian position as a response. Freud ends *Civilisation and its Discontents* with the words: 'And we may well heave a sigh of relief at the thought that is nevertheless vouchsafed to a few to salvage without effort from the whirlpool of their own feelings the deepest truths' (Freud, p.133). Gissing foresees both jadedly and accurately the coming hegemony of emergent mass culture, represented by the success of Kipling's poems and Carnaby's thriving bicycle-factory, producing hundreds of bicycles a week for an increasingly leisured working class. (H.G. Wells, far less hostile to these social developments than Gissing, had chosen a working-class hero with an appetite for bicycling and literature alike as the hero of his novel *The Wheels of Chance*, published the year before.) Rolfe, like Freud, sees a connection between civilization's turning to outdoor activity (such as bicycling) and the prevalence of aggressive drives:

> "Mankind won't stand it much longer, this encroachment of the humane spirit. See the spread of athletics. We must look to our physique, and make ourselves ready. Those Lancashire operatives, laming and killing each other at football, turning a game into a battle. For the milder of us there's golf – an epidemic. Women turn to cricket – tennis is too soft – and tomorrow they'll be bicycling by the thousand; – they must breed a stouter race. We may reasonably hope, old man, to see our boys blown into small bits by the explosive that hasn't got its name yet." (*Whirlpool*, pp.449f.)

Gissing's Masculinities

Gissing's different fictional masculinities constitute an acknowledgement of the hostile and unstable nature of the world his characters inhabit. Rolfe desires to 'toughen up' his son in the advent of the war which threatens his development:

> "The best kind of education would be that which hardened his skin and blunted his sympathies. What right have I to make him sensitive? The thing is, to get through life with as little suffering as possible. What monstrous folly to teach him to wince and cry out at the sufferings of other people! Won't he have enough of his own before he has done? Yet that's what we shall aim at – to cultivate his sympathetic emotions, so that the death of a bird shall make him sad, and the sight of human distress wring his heart. Real kindness would try to make of him a healthy ruffian, with just enough conscience to keep him from crime.... It's natural for a boy to be a good deal of a savage, but our civilisation is doing its best to change that." (p.342)

Gissing wrote to Morley Roberts in 1900 that he would 'rather never see [Walter] again than foresee his marching in ranks, butchering or to be butchered' (Mattheisen et al., vol. 8, p.11; see also p.50, and vol. 7, p.419). Gissing never did see Walter again after 1900. On arriving at manhood, Walter Gissing beat the British drum and was killed in the Battle of the Somme in 1916. (Both Hughie Rolfe and Albert Wager would also have been old enough to fight in the First World War). Alan Sinfield quotes from a letter by another writer's son with even more to prove in the way of his 'manliness' – Cyril, the son of Oscar Wilde:

> The more I thought of [the stain on his name], the more convinced I became that, first and foremost, I must be a *man*. There was to be no cry of decadent artist, of effeminate aesthete, of weak-kneed degenerate. (Sinfield, p.126)

If forced to choose, Gissing would have sided with the 'aesthetes' against the 'sportsmen'. However, in daily life he seems himself to have displayed the performance of a more pronounced masculinity. Perhaps surprisingly, Gissing himself may have embodied the successful union of masculine types which the ending of *The Whirlpool* seems to desire. His pupil Austin Harrison paid tribute both to Gissing's physical strength and athleticism and to his artistic gentleness. Israel Zangwill remembered him in 1896 as a man 'in the flower of his age: a tall strapping Anglo-Saxon, scarcely suggesting the literary recluse' (quoted respectively in Michaux, pp.22ff., and in Halperin, p.236). H.G. Wells, who taught Gissing to ride a bicycle, remembered a Gissing who 'craved to laugh, jest, enjoy, stride along against the wind, shout, "quaff mighty flagons" ' (Wells, vol.2, pp.571f.). The wish that 'something more of repose, of self-possession, and a slightly more intellectual brow, would have made [Carnaby] the best type of conquering, civilising Briton' offers the possibility of a more harmonious masculinity, characterized by extremes neither of nervous bookishness nor of unthinking violence. It is surely one of the characteristics of the unhappy man to wish himself other than he is; still more, perhaps, of the father to wish one's child better or happier than oneself. The presence of a child in the ending of the novel at least allows the hope that nature and civilization might in a future generation be brought into a more comfortable accommodation. Rolfe recognizes that Hughie must, whatever his own nature, adapt to the nature of the world, to a limited extent, thus perhaps avoiding the threats the world has in store for those with an aesthetic temperament:

> "I must send him to the Grammar-School next year," said Rolfe. "It won't do; he must be among boys, and learn to be noisy. Perhaps I have been altogether wrong in teaching him myself. What right has a man to teach, who can't make up his mind on any subject of thought?" (*Whirlpool*, p.452)

Perhaps, in spite of the final estrangement of his father and his namesake, educated in the right way Hugh Rolfe may have the potential to embody the union of the contrasting masculinities hinted at in the novel's frustrated beginning.

Bibliography

Bristow, Joseph (1995), *Effeminate England: Homoerotic Writing after 1885*, Columbia University Press, New York.

Federico, Annette (1991), *Masculine Identity in Hardy and Gissing*, Associated University Presses, London.

Freud, Sigmund (1995; first publ. 1959), 'Civilization and its Discontents', in *The Complete Psychological Works of Sigmund Freud*, 24 vols, trans. by James Strachey, Hogarth Press, London, vol. 21, pp.64-145.

Gatrell, Simon (1982), 'England, Europe and Empire: Hardy, Meredith and Gissing', in Gatrell (ed.), *English Literature and the Wider World, IV: 1876-1918: The Ends of the Earth*, pp.67-82.

Gettmann, Royal (1961), *George Gissing and H.G. Wells: A Record of their Friendship and Correspondence*, Hart-Davis, London.

Gissing, George, ed. Pierre Coustillas (1978; first publ. 1892), *Born in Exile*, Harvester Press, Hassocks (Sussex).

Gissing, George, ed. Pierre Coustillas (1969; first publ. 1886), *Isabel Clarendon*, Harvester Press, Brighton.

Gissing, George, ed. Bernard Bergonzi (1985; first publ. 1891), *New Grub Street*, Penguin, Harmondsworth.

Gissing, George, ed. John Stewart Collis and Pierre Coustillas (1982; first publ. 1903), *The Private Papers of Henry Ryecroft*, Harvester Press, Brighton.

Gissing, George, ed. Patrick Parrinder (1979; first publ. 1897), *The Whirlpool*, Harvester Press, Brighton.

Gissing, George, ed. Pierre Coustillas (1978), *London and the Life of Literature in Late Victorian England: The Diary of George Gissing, Novelist*, Harvester Press, Hassocks (Sussex).

Greenslade, William (1994), *Degeneration, Culture and the Novel 1880-1940*, Cambridge University Press, Cambridge.

Grylls, David (1986), *The Paradox of Gissing*, Allen and Unwin, London.

Halperin, John (1982), *Gissing: A Life in Books*, Oxford University Press, Oxford.

James, Henry, ed. Frank Kermode (1986), *The Figure in the Carpet and Other Stories*, Harmondsworth, Penguin.

James, Simon J. (2003), *Unsettled Accounts: Money and Narrative in the Novels of George Gissing*, London: Anthem Press.

Kipling, Rudyard, ed. M. M. Kaye (1991), *The Complete Verse*, Gollancz, London.

Kramer, David (2000), 'George Gissing and Women's Work: Contextualising the Female Professional', *English Literature in Transition* 43, 3 (2000), pp.316-330.

Mangan, J.A., and James Walvin, eds (1987), *Manliness and Morality: Middle-class Masculinity in Britain and America, 1800-1940*, Manchester University Press, Manchester.

Mattheisen, Paul F., Young, Arthur C. and Coustillas, Pierre (eds.) (1990-1997), *The Collected Letters of George Gissing* (in nine volumes), Ohio University Press, Athens (Ohio).

Michaux, Jean-Pierre (ed.) (1981), *George Gissing: Critical Essays*, Vision, London.

Showalter, Elaine (1987), *The Female Malady: Women, Madness and English Culture 1830-1980*, Virago, London.

Sinfield, Alan (1994), *The Wilde Century: Effeminacy, Oscar Wilde and the Queer Moment*, Cassell, London.

Tosh, John (1999), *A Man's Place: Masculinity and the Middle-Class Home in Victorian England*, Yale University Press, New Haven and London.

Wells, H.G. (1934), *Experiment in Autobiography*, 2 vols., Gollancz, London.

Chapter 7

Whirlpools of Modernity: European Naturalism and the Urban Phantasmagoria

Deborah Parsons

'Was he himself to become a victim of this social disease? Was he, resistless, to be drawn into the muddy whirlpool, to spin round and round amid gibbering phantoms, abandoning himself with a grin of inane conceit, or clutching in desperation at futile hopes?' (*The Whirlpool*, p.201). So Harvey Rolfe contemplates the condition of modernity, at once environment and pathology, in the late nineteenth-century London of George Gissing's *The Whirlpool* (1897). In this engulfing current of restlessness and dis-ease, the social, spatial and subjective tumult of modern life, the 'whirlpool of high capitalism', 'the whirlpool of metropolitan culture' and the 'whirlpool of modernity', as John Goode notes, are inextricably linked within the 'phantasmal' world of the metropolis (Goode, p.183). Of all Gissing's studies of the estrangement of human life by social and economic forces, it is *The Whirlpool* that, in its recognition of the new rhythms of metropolitan modernity, most extends and complicates literary naturalism's pseudo-scientific claim to empirical exactitude.

Images of the unstable energies of bourgeois modernity abound throughout naturalist fiction; financial speculation and the scramble for the spoils of expansion, the profusion of goods and equally abundant refuse of consumerism, the inflation of monetary value, fortunes and desires. Drawing on Georg Simmel's theories of metropolitan consciousness, this paper focuses on the representation of the experience of urban modernity in novels by Gissing, Emile Zola and Benito Peréz Galdós. I argue that despite their typically broadly positivist position, recording the struggle for survival within the modern city and picturing those who thrive upon and those who succumb to its influences, all three writers ultimately eschew a crudely deterministic scientific ideology in their ambivalent fascination with the phantasmagoria of urban modernity. Of course late nineteenth-century London, Paris and Madrid in many ways present distinct social and cultural cityscapes, at different stages of imperial might and spectacular modernity. Yet for Gissing, Zola and Galdós respectively, each concentrates the social, economic and psychological climate of the burgeoning European metropolis. Declaring with more or less vigour the social-scientific focus of their works, each took up the task of portraying the multi-levelled middle classes within the social and economic landscape of metropolitan modernity. Galdós argued in 1870, for example, that

> the social order nowadays is built on the middle-class: through its initiative and intelligence it has taken on the sovereign role in all nations; it is there that nineteenth century man is to be found, with all his virtues and vices, his noble, insatiable aspirations, his passion for reforms, his frantic activity. (Labanyi, pp..32f.)

Just as Zola's Rougon-Macqart novels (1871-1893) depict the transformation of Paris into Benjamin's 'capital of modernity', so Gissing's 1890s novels record the social and financial complexities of late-nineteenth-century London, and Galdós' *novelas contemporaneas* (contemporary novels) the social, political and spatial development of bourgeois Madrid from the fall of Isabel II in 1868 to the Restoration of the Bourbon monarchy in 1875. All three writers embrace the naturalist aesthetic in their depiction of an organic urban milieu, their social-scientific focus on the influence of heredity and environment, and their relation of personal histories to the broader social and political climate of city and nation. Turning from labour to leisure in their depiction of bourgeois modernity, however, and from spaces of production to those of consumption, they also reveal the challenge to objective observation within the modern metropolis, and the illusions and distorted perceptions that pervade its everyday life. Through the perceptions and experiences of their female protagonists, in particular, the novels discussed here trouble the sociological gaze of scientific realism as they mirror the artifice of the modern city.

The lack of something definite

The aim of the artist of modern life, according to Charles Baudelaire's often quoted essay of 1863, was to capture the energies of modernity, the 'transient, the fleeting, the contingent' nature of existence (Baudelaire, p.403). Baudelaire recognized in the new landscape of the metropolis a challenge to the conventions of realism, and the need for a new aesthetic idiom of everyday life. The 'age of high capitalism', as Walter Benjamin declares in his study of Baudelaire, brought with it the phantasmagoria of modernity. T.J. Clark describes the Paris created by Baron Georges Haussmann, Napoleon III's Prefect of the Seine, as 'characterized by shifts and disguises, by too many surfaces and too few lines of demarcation' (Clark, p.47). Its landscape is full of 'places laid on for display but also for equivocation; places where people are hard to make out, their gestures and expressions unconvincing, their purposes obscure' (p.48). It was this social ambiguity of the new spaces of urban modernity more generally, and the inscrutability of the urban crowd that they give rise to, Benjamin suggests, that resulted in the urgent impulse to categorization and differentiation on the part of nineteenth-century writers and social commentators. The very instability of social boundaries in the modern city of speculation and spectacle, however, where fortunes rise and fall according to the 'securities' of the stock market, and the appearance of status can be bought on credit, resists such attempts.

Naturalism and impressionism, the former attempting to systematically document what it observes, the latter to represent how it observes, were related forms of response to the unstable modern world of speculative finance and spectacular consumption. Both take as their focus the modern urban environment, in particular the spaces in which the social elite, bourgeois and working classes, and demi-monde coalesce, and in which the boundaries between them become vague. Moreover, each advocates a realism of everyday life, in which modern art is that which renders the exterior world through the perception of the artist. Extending naturalism's aim at the accurate social observation of everyday life, however, the impressionist impulse was to record the subjective experience of modernity; its rhythms, sense impressions and distorted points of view. As Jules-Antoine Castagnary defined the artists of the first Independent Exhibition in Paris in 1874, 'they are impressionists in that they render not the landscape but the sensation produced by it' (see Shiff, p.388).

Conceptualizations and critiques of modernity by the new social-scientific disciplines of the late nineteenth century typically focus on the impact of the spatial and temporal 'shocks' and discontinuities of the modern environment on the physical and mental life of its inhabitants. For the sociologist Georg Simmel, the constantly bombarding stimuli of the metropolis (which Simmel highlighted as the key site of modernity and its consumer capitalist economy) resulted in a heightened yet delicate sensory capacity on the part of the modern urban dweller. As he delineates in his 1903 essay on 'The Metropolis and Mental Life', the objective world of modernity, characterized by 'the rapid crowding of changing images, the sharp discontinuity in the grasp of a single glance, and the unexpectedness of onrushing impressions', so overwhelms the receptive capacity of the individual that the mind responds with a defensive mechanism of subjective dissociation (Frisby and Featherstone, p. 175). Thus the characteristic mentality produced by the modern environment is for Simmel the blasé attitude, in which the individual becomes indifferent to the multiple stimuli that surround him. In an early outline of this theory in 1890, Simmel had drawn explicitly on the muted hues of the impressionist palette to describe the sense experiences of urban dwellers, whose nervous reactions 'have become increasingly sensitive, so that instead of strong colours and their contrasts they can only cope with the pale, semi-faded tints' (Frisby, 2001, p.112). He returns to them again in what was perhaps his major work *The Philosophy of Money* (1900), in order to render the impressions of the blasé urbanite, who 'experiences all things as being of an equally dull and grey hue, as not worth getting excited about' (Simmel, 1990, p. 256). The blasé sensibility still demands the stimulations to which it has grown accustomed but, in a world in which all has been reduced to the values of circulation and exchange, its restless desires can never be satiated. 'The lack of something definite at the centre of the soul', Simmel states, 'impels us to search for momentary satisfaction in ever-new stimulation, sensation and external activities. Thus it is we become entangled in the instability and helplessness that manifests itself as the tumult of the metropolis' (p.484).

In his analysis of the psychological effects of capitalist metropolitan culture, Simmel had little sympathy for the scientific affectation of what he described as a 'mechanistic naturalism'. He developed a sociology of modernity that emphasized the fluid and fleeting phenomena that he recognized as characterizing the social interactions of everyday life. His assertion in his 1896 essay 'Sociological Aesthetics' that 'the essence of aesthetic observation and interpretation lies in the fact that the typical is to be found in what is unique, the law-like in what is fortuitous, the essence of things in the superficial and transitory' (Frisby, 2001, p.103) recalls Baudelaire's definition three decades previously, and acknowledges the importance of the phantasmagoria in the study of the culture of high capitalism. Stimulation, desire, consumption, indifference: these are the psychological conditions produced by the circulatory rhythm of money in modern society. The analysis of the mature money economy thus entails for Simmel an approach that registers both the objective and the subjective culture of metropolitan modernity, a combination of the ethnographic and the phenomenological that Frisby has argued might be seen as a methodology of 'sociological impressionism' (Frisby, 1991).

Naturalism and the phantasmagoria

Despite Zola's strident claims to a somewhat paradoxical literary stance of scientific objectivity, often in his fiction (as in the late nineteenth-century novels of Gissing and Galdós) there are moments when social observation leads to an acknowledgement of and engagement with the shifting discontinuities and superficialities of metropolitan modernity. Raymond Williams suggests that in Gissing's portrayal of the big city experience, the familiar figure of the detached, realist urban observer 'is overwhelmed by the crowds and the ugliness', and that in his description of Dickens' London as 'a marvellous rendering of the impression received by any imaginative person who, in low spirits, has had occasion to wander about [its] streets', he effectively describes his own (Williams, 1973, p.235). Similarly, it is often at the point of Zola's most overloaded descriptions of 'things' that both protagonists and readers feel most removed from actuality. Naturalism too, in other words, when faced with a glut of accumulating sense impressions, like Simmel's over-stimulated urbanite reaches surfeit. Haussmann's Paris was regularly described by its contemporary critics, as Clark reveals, as a 'parade, phantasmagoria, dream, dumbshow, mirage, masquerade' (Clark, p.66). Such metaphors of 'visual untruth' also pervade Gissing's and Galdós' later representations of metropolitan culture. Three years before his depiction of the 'phantoms' of *The Whirlpool*, for example, in *In the Year of Jubilee* (1894), Gissing portrays central London as a place of shifting façades; in its ostentatious parade of a diminishing imperial might as much as its ever-changing, advertisement-covered surfaces. The city is dominated by the swarm of its faceless crowds and the veneer of advertisements that layer any available surface – for instance, at King's Cross underground station:

> High and low, on every available yard of wall, advertisements clamoured to the eye: theatres, journals, soaps, medicines, concerts, furniture, wines, prayer-meetings – all the produce and refuse of civilisation announced in staring letters, in daubed effigies, base, paltry, grotesque. A battle-ground of advertisements, fitly chosen amid

> subterranean din and reek; a symbol to the gaze of that relentless warfare which ceases not, night and day, in the world above. (*Jubilee*, p.309)

Squeezed into a tramcar on a journey to Westminster and the Jubilee celebrations, Nancy Lord hardly registers these everyday clashing stimulations of the metropolis. Idly reading the advertisements pasted above the passengers' heads, like Simmel's blasé urbanite she remains indifferent to their import:

> Somebody's "Blue;" somebody's "Soap;" somebody's "High-class Jams;" and behold, inserted beneath the Soap and the Jam – "God so loved the world, that He gave His only Son, that whoso believeth in Him should not perish, but have everlasting life." Nancy perused the passage without perception of incongruity, without emotion of any kind.... the associations of the market-place had in this respect blunted her sensibilities. (pp.60f.)

Indeed it is not until Nancy escapes from her companions into the vast crowd surging towards Regent Street, feeling its physical contact and breathing its 'atmosphere of exhalations', that she is moved to thoughtless and 'vulgar abandonment' by the intensity of the distractions that surround her (p.69). It is perhaps appropriate that Nancy's guide to the city in the first part of the novel, who twice saves her from the whirlpool of the crowd in which she quickly becomes disoriented, is the brazen yet enterprising advertising agent Luckworth Crewe. Inspired not by the London scene's 'historic associations' (p.106), but rather by its stories of speculation, thriving business and inflating rents, Crewe regards the cityscape in terms of its marketable potential and capacity for commercial profit. Astutely and energetically versed in the ideology of the modern economy, it is Crewe and his female counterpart, the vulgar but determinedly independent Beatrice French (whose Fashionable Dress Supply Association similarly capitalizes on the exchange value of the image), who ultimately represent the 'Age of Progress' that the Jubilee celebrates. Crewe survives and profits by his understanding of the modern phantasmagoria, and it is perhaps a sign of Gissing's refusal of the gaudy commercialism that Crewe embodies that his colourful yet disconcerting presence all but disappears from the second half of the novel, Gissing returning his attentions to his fantasy of an unorthodox domestic idyll between Nancy and the rather ill-defined figure of Lionel Tarrant.

Galdós goes even further in his depiction of bourgeois Madrid as a site of uncertainty, fantasy, and disillusionment, in which reality and fiction become indistinct. *Fortunata y Jacinta* (1886), generally regarded as Galdós' most studied example of sociological realism, portrays the history of the development of Madrid from a provincial town and the seat of a collapsing empire in the first decades of the nineteenth century, into a bourgeois consumer metropolis by those of its end. The successful rise of the Santa Cruz family, from ownership of a modest fabric shop to leading members of Madrid's mercantile business aristocracy, represents in microcosm this burgeoning development, which coincided with the growing financial power of the middle class, and the creation of the social and physical landscape of a secular city of modernity:

> [It] was the era when the middle class was beginning to come into its own, taking up the jobs created by the new political and administrative system, buying up all the properties (of which the Church had held the title) with instalment plans, thus making itself the chief landowner and beneficiary of the budget; in a word, gathering the spoils of absolutism and clericalism to found the empire of the frock coat. (*Fortunata y Jacinta*, p.28)

For Galdós the craft of fictional realism paralleled the illusory nature of Madrid's own claim to metropolitan modernity, however, and he questions the transparent representation of its bourgeois society even as he represents and affirms it. His description of the biography of the Santa Cruz family in an early manuscript of *Fortunata y Jacinta*, for example, as a 'simulacro' or 'replica' of the urban environment he recreates (Parsons, 2003, p.37), is in part an assertion of a realist aesthetic, but paradoxically also hints at undercutting that aesthetic; reflecting the instability of a society built on socioeconomic exchange and the false values of a middle class constantly mortgaging itself to a material status and economic power that it does not possess.

Hyperaesthesia and the female subject

The modern city's 'immense abundance' of products, influences and interests, Simmel argues, results in 'endless habits, endless distractions and endless superficial needs' through which the modern urban dweller becomes estranged from his 'most distinctive and essential being' (Simmel, 1990, pp.483f.). The cyclical or reciprocal logic of modernity that is here implied – a profusion of stimuli leads to a demand for ever-increasing, ever-new stimuli and so forth – links inextricably the socio-economic and psychological landscapes of modernity. Significantly, while it is through the activities of male characters that Gissing, Zola and Galdós generally depict the objective operations and profits of financial exchange in the modern economy, it is typically through the impressions of young female protagonists that each writer registers the inner life of the modern phantasmagoria. Mass cultural modernity, Andreas Huyssen argues, has been consistently gendered as feminine by Western discourse, 'traditionally described as the threat of losing oneself in dreams and delusions and of merely consuming rather than producing' (Huyssen, 1986, p.47). Gissing, Zola and Galdós frequently portray the mental life of modernity through young female protagonists, exploring through the fantasies and distorted visions of Alma Rolfe in *The Whirlpool* and Nancy Lord in *In the Year of Jubilee* (1894), Renée Saccard in *La Curée* (*The Kill*, 1872), or Isidora Rufete in *La desheredada* (*The Disinherited*, 1881), for example, the frenzied, inexorable desires and states of hypersensitivity, neurasthenia and ennui that Simmel identifies with the mental life of the reified world of modernity.

If Zola's exhaustive detailing of the speculative dealings and accoutrements of new wealth of Aristide Saccard in *The Kill* exemplifies his naturalist eye, it is to an impressionist style that he turns in order to express the restless desires of Saccard's bored, spendthrift wife Renée. The spectacle of the Bois de Boulogne under an October dusk that opens the novel, a three page set piece of Manet-inspired imagery, is a blur of fading light and shade on colour:

> This scrap of Nature, that seemed like a newly painted piece of scenery, lay bathed in a faint shadow, in a pale blue vapour which succeeded in lending to the background an exquisite charm, an air of entrancing artificiality. On the other bank, the Châlet des Iles, as though newly varnished, shone like an unused toy; and the paths of yellow sand, the narrow garden walks that wind among the lawns and run along the lake, edged with iron hoops in imitation of rustic woodwork, stood out more curiously, in this last hour of daylight, against the softened green of grass and water. (*The Kill*, p.8)

Nature makes way for spectacle and artifice in this worldly Parisian leisure-ground, where lady and courtesan, gentleman and industrialist intermingle amidst gleaming reflective surfaces. Renée, however, is '[a]ccustomed to the ingenious charms of this perspective' and lies back listlessly and unseeingly in her carriage. Her myopia, Zola implies, is not merely a physical complaint but the result of the indifference of the blasé urbanite, of her 'sick heart and weary flesh' (p.13). Casting at the bourgeois scene 'the despairing glance of women who do not know in what form of amusement to take refuge', her eyes glow instead with 'the ardent desire of unsatisfied curiosity' for 'something different.... nameless enjoyment' (pp.10-13).

One of the psychological effects of the money economy, Simmel notes, is that '[the] mere increase in the supply of money that one has in one's hand intensifies – quite regardless of any awareness of its mere relativity – the temptation to spend money, and in so doing promotes a greater turnover in commodities, an increase, acceleration and multiplication in economic conceptions' (Simmel, 1990, p.499). Thus Renée, 'in mad intoxication, fill[s] Paris with the clatter of her equipages, the sparkle of her diamonds, the vertigo of her riotous existence' (*The Kill*, p.95), while her husband pursues ever riskier speculative investments. Their mansion on the new boulevard Malesherbes, its function the ostentatious display of Saccard's otherwise intangible funds, is compared to 'the purse-proud, foolish importance of a female parvenu', its windows 'sheets of plate glass so wide and so clear that they seemed like the window-fronts of a big modern shop, arranged so as to display to the outer world the wealth within' (p.19). Yet each whirl of ever-increasing spending and ever more reckless gratification of her sensory lusts is followed for Renée by a renewed tedium and the return of a restless hunger for further excitement.

'As long as we are not yet in a position to buy things,' Simmel observes, 'they affect us with their particular distinctive charms. Yet as soon as we easily acquire them with our money, those charms fade away.' 'The search for mere stimuli in themselves,' he continues, 'is the consequence of the increasing blasé attitude through which natural excitement increasingly disappears', the result of the collapsing of qualitative value into the quantitative exchange values of the modern money economy (Simmel, 1990, p.257). Thus Renée contemplates with languor and impatience 'her past life, the instantaneous satisfaction of her appetites, the fulsomeness of luxury, the appalling monotony', desperate for 'something else, something that would happen to nobody but one's self, that would not be met with every day, that would give a rare, unknown enjoyment' (*The Kill*, pp.15, 14). Somewhat paradoxically, the blasé attitude is therefore often co-existent with the contrasting state of neurasthenia, characterized by an intensity of nervous life in response to the material stimuli of the metropolis, a desire that is never satiated, despite 'ever-new stimulations, sensations and external activities' (Simmel, 1990, p.484).

Lacking the purchasing power of Renée Saccard, Isidora Rufete in Galdós' *The Disinherited* remains in thrall to the excitements of consumption, caught within the whirlpool of circulation and exchange of the commodity. Galdós' contemporary novels regularly testify to the existence of an incipient yet vibrant Madrileñian consumer culture, in which shopping was a seductive and addictive activity. Indeed, although Zola's *Au Bonheur des Dames* (1883) offers perhaps the most notorious dramatization of material desire as the irresistible seducer of women through the vast, spectacular space of the department store, *The Disinherited*, written two years before, already presents shopping as an activity in which women were seen to indulge with sensual and irrational pleasure. At the beginning of chapter seven of the novel, Isidora takes a long walk through the commercial heart of Madrid, under the pretext of a journey to mass. The city for Isidora is ultimately a commercial pleasure zone in which she can wander and consume with freedom and delight. 'Looking into shop windows, attracted by anything pretty, rich or sumptuous', she revels in their infinite bazaar of opulent commodities –

> rich furs, endless fine clothes, ties, trinkets that enchant the eye with their strangeness, objects that combine invention and patient industry, putting to use gold, silver, nickel, Russian leather, celluloid, cornelian, jet, amber, brass, rubber, coral, steel, satin, glass, talc, mother-of-pearl, shagreen, porcelain and even horn.... exotic furniture inlaid with ivory, oak carved in feudal style, inlaid walnut, majestic double beds and finally, bronzes, ceramics, clocks, cruets, candelabra. (*The Disinherited*, p.111)

Entitled 'Taking possession of Madrid', the chapter depicts Isidora's limitless material desires with an appropriately materialist inventory, yet it also reveals the irony of her aspirations to this city of material wealth and comfortable leisure.

Far from a wealthy consumer, Isidora is in fact a provincial parvenu from La Mancha, who believes herself to be the illegitimate daughter of a noblewoman and who has arrived in the city to demand recognition and inheritance. Her voracious reading of romantic serial novels, and uncritical acceptance of their rags-to-riches tales of beautiful young women, have convinced her that her claim is bound to succeed. She has internalized the image found in the popular press of the high-society woman as fashionable and leisured consumer, and her vanity and self-importance incline her towards material and sensual pleasure. The goods on display in the shop windows, described by the narrator as 'gaudy superfluous rubbish' (p.111), glitter enticingly before Isidora's fantasizing eyes, the reflection of her body in the glass seemingly adorned with the goods on display. In her vanity and self-absorption she creates herself in terms of objects, engaging in an act of self-commodification for her own consumer fantasies. As Rachel Bowlby notes in her study of consumer culture in the naturalist novel, '[t]hrough the glass, the woman sees what she wants and what she wants to be' (Bowlby, 1985, p.32). Isidora's devouring gaze quickly turns into overwhelming desire: 'She had to buy something, a little something' (*The Disinherited*, p.111). The purchase of a parasol is rapidly followed by gloves, a fan, a brooch, earrings, a purse, perfume, hair-pins, a comb, embossed note-paper, as the limits of Isidora's pocket are overcome by an insatiable appetite for extravagant commodities.

In Isidora Rufete, Galdós created a modern female protagonist who may be conceited and shallow in her material wants, but who also becomes skilfully conversant in, and indeed identifies herself through, the rhetoric of spectacular consumerist urbanity, the reflections of the phantasmagoria. Shortly after arriving in Madrid, her social vanity already evident, Isidora takes a Sunday excursion to the Prado Museum and Retiro Gardens with the medical student Augusto Miquis. Here her attention is quickly caught by the contrast between her unsophisticated appearance and the spectacle of wealthy society around her. As they walk home, they pass the elegant urban artery of the Paseo de la Castellana, the city's most fashionable avenue, surging with elite society on display, a parade of carriages that Miquis refers to in disgust as 'a torrent of vanity' (p.71). Later in the novel, as eminent doctor and Galdós' voice of institutional middle class society, he will assume the role of Isidora's moral custodian, and it is thus ironic that it is he who now explains to her the principles of imitation and illusion on which she will base her future:

> "You'll see all social classes here. They come to be observed, to be compared and to see the respective distances between each other, in order to prepare for the assault. It's a case of climbing up the next rung on the ladder.... They all rub shoulders and are tolerant of each other because equality rules over all. Nobody envies an illustrious name, but they do envy material possessions. As each one is furiously determined to go up in the world, he begins by posing. These improvisations stimulate the appetite." (p.74)

Exhilarated by the spectacle of the Castellana, Isidora ignores the implicit critique in Miquis' commentary, and responding only to the ostentatious display of the fashionable nobility, she is overwhelmed by the city's incandescent allure:

> Her whole heart and mind were swept along on the swell. Far in the distance quite a few gas lights were shining through the dust of the Prado. That mist that forms from the breath of the city population, the evaporation from sprinkling water and the continual sweeping (the trains of dresses acting as brooms), was being illuminated into a fantastic glow, like a luminous radiation from the ground itself. (p.75)

In his description of the pseudo-luxury of the Castellana, Miquis thus unwittingly hands Isidora what is effectively a conduct manual. The cause of her subsequent overspending is not merely a greedy desire for possessions, but more fundamentally a desire for identity.

Rather like the new city itself, springing up from the dust of demolition and expansion, Isidora has detached herself from her provincial past, yet, at the same time, her new sense of self requires consolidation. Depending on material appearance for her sense of self, and on the narratives of mass consumer culture for her attitudes and actions, the goods she buys are the trivial luxuries of fashion that symbolize and convey the outward impression of a social status that she does not possess. Bowlby argues that in the world of consumer capitalism, '[t]he commodity makes the person and the person is, if not for sale, then an object whose value or status can be read off with accuracy in terms of the things he has and the behavioural codes he adopts' (Bowlby, 1985, p.12). In a society in which commodities and the principle of spectacle become constitutive of identity, Isidora consumes in order to create herself in the image of the wealthy women she sees on the Castellana and indeed believes herself to be. Yet although she excels in exactly this approximation of identity through commercial accumulation, in her desperate need for the objects that will sustain that identity she ultimately becomes a figure who is indeed herself for sale. The street for Isidora is the space of both her dreams and her disillusionment.

Galdós' longest, concentrated dramatization of Isidora's subjective experience of Madrid is in her frenzied disorientation after the Marquesa de Aransis rejects her claim to aristocratic inheritance. Walking into a city that is suddenly unfamiliar to her, turned to rubble by the contradiction of her fantasies, her confused mind perceives that there has been a '[t]otal transformation. The world was quite different. Nature itself, the air and the light, all quite different' (*The Disinherited*, p.213). Galdós represents her increasingly hypersensitive response to the city as she moves aimlessly, then meditatively, and then excitedly through its streets. Her trajectory at first takes her from the Aransis palace to the Plaza de Oriente, the city of the seventeenth-century Court in which, her grand presumptions now destroyed, she no longer belongs. Isidora's desire for nobility, however, is never really much more than a desire for access to the pleasures of urban modernity, and, turning away from the aristocratic society of old Madrid that has denounced her, she walks towards the Puerta del Sol, the city's irrepressibly energetic central square: 'As she drew closer to the centre of Madrid and felt the throbbing vitality of the city, the idea of living grew even stronger within her: she wanted to experience, to see, to taste' (p.218). As she moves quickly to where '[t]he Puerta del Sol, beating like a heart permanently stimulated, revealed its fast moving breathless life', her excitement increases:

> Madrid, at eight thirty in the evening, is pure enchantment, an open market, a display of delights and endless pleasures. The theatres attract with their gas lit signs, the shops draw attention with the persuasive humbug of the window displays, the cafés fascinate with their buzz of conversation and casual atmosphere, where sweet idleness and gossip flourish. Wandering about at this time of the evening has all the attractions of a stroll and the seductions of a journey into the unknown. (p.219)

The mass of people is greater than usual, and Isidora tingles with postive energy, 'excited by the endless repeated impressions of the jostling, brushing, pushing crowd, of things seen and desired' (p.220). The Republic has been declared, and the city suddenly seems a place of possibility. Caught up both physically and emotionally by the crowd surging towards the Congreso on the Carrera de San Jerónimo, she exclaims at 'How many men there were, and women too!', exhilarated by '[c]ontact with the masses, that fluid magnetic conductor of mysterious desires which flows from one body to another' (p.221).

The Puerta del Sol is an ambiguous site, however, at once a public arena where the new Republic is being shouted from one man to another, and a phantasmagoria of 'enchantment', 'display' and 'idleness' by which the frivolous Isidora is easily seduced. It is Carnival time, and the shop windows within which she has constantly imagined her reflection are thus full of 'masks, disguises and carnival heads.... a façade of leering grimaces' (p.220). The eve of the Republic is as portentous of folly as it is of equality. Isidora's immediate act is to seek out the affections of the wealthy Joaquín Pez and the promise of quick personal advancement, a prediction of the similar disintegration of the new Republic in its self-interested seizing of economic and political spoils. Galdós' modern Madrid is a city that consumes voraciously yet produces almost nothing, obsessed only with material appearance and social status, and Isidora is presented as its analogue. His narrative of her desire, extravagance and fall in the Spanish capital, however, while according with a naturalist passion for over-determined endings, offers a compelling representation of the mental life of the metropolis.

Renée Saccard and Isidora Rufete seek the fulfilment of their nervous urban desires in material consumption, lusts and fantasies, their increasingly overwhelming conditions of ennui and neurasthenia the result of the phantasmagoria of the capitalist environment. In *The Whirlpool*, the neurasthenic's defensive rejection of the materiality of the metropolis leads to the aestheticist detachment into which Harvey Rolfe has retreated in his bachelordom at the beginning of the novel, and at its extreme to the 'cold abstraction' (*Whirlpool*, p.34) of a Cyrus Redgrave or a Sibyl Carnaby. Alma Frothingham, by contrast, nurtured in the unstable domestic environment provided by the legacy of her psychologically fragile mother and the financial speculations of her banker father, is far less adept at surviving the struggle of modernity. Yet Alma's hereditary nervous sensibility, given to sudden, capricious enthusiasms, leaves her particularly susceptible to 'the whirlpool way of life' (p.147). The imagined idyll of domestic rural simplicity and motherhood with Rolfe in Wales soon leaves her bored, rebellious and increasingly hysterical, and reawakens her delusions of a career as a professional musician. Relocating to the 'healthily' distant London suburb of Pinner, Rolfe hopes, only half an hour from Baker Street, will allow her 'the precious advantage of being able to use London for all legitimate purposes, without danger of being drawn into the vortex of its idle temptations' (p.178). The influence of the city reaches out beyond its boundaries, however. Alma's nervous temperament quickly responds, and the 'cool, lymphatic flow' (p.212) with which Gissing at one point rather surprisingly endows her is soon flooded by the fantasies and fears of an excitable imagination. After the initial exuberance at returning to London, her journeys to the West End leave her suffering from debilitating headaches, the symptoms of 'some nervous disorder, not to be shaken off by any effort' (p.244). Exhausted by the whirlpool, she yet works feverishly toward her rash public recital, turning to the shady influence and sycophancy of a decadent demi-monde for the gratification of her vanity until she finds herself increasingly embroiled within a web of manipulation and deceit beyond her control. Her over-excited and over-taxed nerves generating the pathological condition defined by Simmel in 1890 as 'hyperaesthesia, the unhealthily accentuated impact of every impression' (Frisby, 2001, p.113), Alma sinks into a delirium of mental illness and morphine-induced hallucinations as London society becomes for her increasingly 'unreal' (*Whirlpool*, p.284). Easy prey for the urbane acuity of the coolly calculating Sibyl, Alma's death is the result of her neurasthenic maladjustment to the tumultuous energies of the modern metropolis.

In an account that echoes Gissing's image of the whirlpool of modernity, Simmel notes:

> The specialization of our times produces the rush from one impression to the other, the impatience for enjoyment, the problematical strivings to compress together in the shortest possible time the largest possible sum of acquisitions, interests and enjoyments. The colourfulness of metropolitan life, both on the street and in the drawing room, is both the cause and the consequence of this continuous striving. (Frisby, 2001, p.103)

Conventionally recognized for their scientific realism, Gissing, Zola and Galdós, like Simmel, I suggest, are led by an initially socio-economic focus on the circulation and exchange values of capitalism, its speculation and spectacle, to an acknowledgement and investigation of these nervous energies of the metropolis. Through the subjectivities of their female protagonists in particular, moreover, they register the new and pathological experiential modes of modernity. Disorientating rather than confirming the sociological gaze, the phantasmagoria of metropolitan modernity reflects back to the observing eye its illusions and artifice.

Bibliography

Baudelaire, Charles, trans. P.E. Charvet (1981), *Selected Writings on Art and Artists*, Cambridge University Press, Cambridge.

Benjamin, Walter (1973), *Charles Baudelaire: A Lyric Poet in the Era of High Capitalism*, New Left Books, London.

Bowlby, Rachel (1985), *Just Looking: Consumer Culture in Dreiser, Gissing and Zola*, Methuen, London and New York.

Clark, T.J. (second edn., 1990), *The Painting of Modern Life: Paris in the Art of Manet and His Followers*, Thames and Hudson, London.

Frisby, David (second edn., 1991), *Sociological Impressionism: A Reassessment of Georg Simmel's Social Theory*, Routledge, London.

Frisby, David and Featherstone, Mike (eds.) (1997), *Simmel on Culture: Selected Writings*, Sage, London.

Frisby, David (2001), *Cityscapes of Modernity*, Polity, Cambridge.

Galdós, Benito Pérez, ed. and trans. Lester Clark (1976; first publ. as *La Desheredada*, 1881), *The Disinherited*, Phoenix House, London.

Galdós, Benito Pérez, ed. and trans. Agnes Moncy Gullón (1986; first publ. 1886), *Fortunata y Jacinta*, Penguin, London.

Gissing, George (1997; first publ. 1894), *In The Year of Jubilee*, Harvester Press, Hassocks.

Gissing, George (1997; first publ. 1897]), *The Whirlpool*, Everyman, London.

Goode, John (1978), *George Gissing: Ideology and Fiction*, Vision Press, London.

Huyssen, Andreas (1986), *After the Great Divide: Modernism, Mass Culture, Postmodernism*, Indiana University Press, Bloomington.

Labanyi, Jo (ed.) (1993), *Galdós*, Longman, London.

Parsons, Deborah (2003), *A Cultural History of Madrid*, Berg, Oxford.

Shiff, Richard (1978), 'The End of Impressionism', *Art Quarterly*, 1:4.

Simmel, George (1990), *The Philosophy of Money*, Routledge, London.

Williams, Raymond (1973), *The Country and the City*, Oxford University Press, Oxford.

Zola, Émile, trans. A.Teixeira de Mattos (1967; first publ. as *La Curée*, 1872), *The Kill*, Arrow, London.

Chapter 8

'To show a man of letters': Gissing, Cultural Authority and Literary Modernism

Martin Ryle

Gissing's study of Dickens, written near the end of his career, can be read in some respects as an implicit vindication of his own practice. Like his novels, it is marked by contradictions which give it much of its interest. Some passages suggest that Gissing rather envies the easy relation Dickens enjoyed with his readers. At the outset, he notes that Dickens' 'years of apprenticeship to life and literature were those which saw the rise and establishment of the Middle Class, commonly called "Great" ' (*Charles Dickens*, p.2). Shared tastes and sentiments kept Dickens close to this growing middle-class audience, making him 'the mouthpiece of his kind, in all that relates to simple emotions and homely thought' (pp.77f; cf. pp.34 and 175f.). 'Mouthpiece' is, however, an equivocal term (and one we shall encounter again): Dickens' closeness to his readers may have given him a certain representative authority, but his ready acquiescence in their world-view is identified as a damaging limitation.

I start by bringing together some comments on this theme from the Dickens essay, and suggesting their relevance to Gissing's fiction. Gissing often implies that authentic writing depends on the adoption of an antagonistic, embattled stance vis à vis the public. I consider how that kind of writerly authority is represented and enacted in some of Gissing's novels, my main focus being on writer-protagonists, from Osmond Waymark to Henry Ryecroft. I seek in particular to show that for Gissing, the writer's intelligence (or 'brains') involves a resented but ineluctable engagement with 'the social question'.

In my conclusion, I turn to broader cultural themes and discuss some ways in which Gissing anticipates literary modernism. Modernist writers and their proto-modernist forerunners also took an ambivalent stance towards their public. In the late Victorian period, novelists who sought formal and stylistic originality, devoting themselves to what Gissing called 'our grave Art of Fiction, a bitter task-mistress' (p.43), found they were writing for a select audience. In his pioneering study *The Uses of Obscurity*, Allon White noted that James, Meredith and Conrad were repeatedly mentioned as appealing to an élite readership (White, p.32). Later, almost all the movements of the modernist avant-garde 'claimed to be anti-bourgeois' (Williams, 1989, p.53). Generally, however, the bourgeoisie 'provided the only real audience [modernist writers] could expect to have' (Nicholls, p.14).

Gissing registers the cultural climate in which anglophone literary modernism grew, charting new social formations and their impact on the writer's consciousness and practice. He is an acute observer of the fragmentation of the literary audience at the *fin de siècle*, and of its consequences. *New Grub Street* offers a classic statement of the opposition between commercial and literary value on which modernism's self-conception was in part founded, and depicts (as does *In The Year of Jubilee*) the cultural marketplaces in and against which it emerged (see Huyssen, pp.18ff.; Bowlby, pp.8f.). However, Gissing in his writing is no proto-modernist. His fiction is quite unimaginative formally. Its conventions of narration, commentary and character-drawing are derivative, with little evidence of deliberate rethinking. His books engage us despite, and even because of, their aesthetic limitations and compromises. They thus cast an interesting light on the work of those modernists who dealt with the antagonism between writer-intellectual and audience by encoding in new formal and aesthetic practices their resolution of what Gissing had felt to be a social and political problem.

To show a man of letters

For Gissing, the possibility of easy relations between author and audience depends both on social-historical trends and on the author's temperament and principles. It was not just that circumstances aligned Dickens' personal history with that of the rising middle class. With a readiness that to some novelists would be inconceivable, he deferred to the tastes and susceptibilities of his readers, feeling this as no constraint.

> To write a novel in a spirit of antagonism to all but a very few of his countrymen would have seemed to him a sort of practical *bull*; is it not the law of novel-writing, first and foremost, that one shall aim at pleasing as many people as possible? (*Charles Dickens*, pp.61f.)

The reverse image is clearly legible, here, of the writer whose 'spirit of antagonism' drives his work; and for whom the aim of 'pleasing as many people as possible' is improper, even contemptible, since he feels in sympathy only with 'a very few'. A little later, noting that 'no distinctly intellectual person figures in [Dickens'] books', Gissing points out that we can hardly regard David Copperfield as an exception: for 'who really thinks of David as a literary man?... Had Dickens intended to show us a man of letters, he would have failed most grievously; of course he aimed at no such thing; the attempt would have cost him half his public' (pp.95f.). Again, the reverse of this gives Gissing's own practice, where representations of intellectuals and 'men of letters' recur, notably the writer-protagonists of *The Unclassed*, *New Grub Street*, *In the Year of Jubilee* and *The Private Papers of Henry Ryecroft*. Apart from their documentary interest as depictions of the trade of letters (Bowlby, pp.95-117; Williams, 1987, pp.172ff.), these remind the reader that culture is made and used within a network of economic and ideological determinations – potentially alienating that 'half' of the public that prefers a less self-conscious kind of reading. They also show the attempts of 'men of letters' to sustain some form of authority in a world where writers may feel, with Ryecroft, that 'literary work' done for the marketplace has little or nothing to do with 'literature' as conventionally esteemed (*Ryecroft*, pp.84f.).

Other comments in *Charles Dickens* underline Dickens' failure to value intellect. Gissing notes that unlike Kingsley in *Alton Locke*, he portrays no working-man distinguished by intellectual attainment, and seems to regard the moral qualities he values as all the more valuable when found in the uneducated (*Charles Dickens*, pp.202f; cf. pp.19, 205f.). Stephen Blackpool in *Hard Times*, his sole attempt to show 'the workman at war with capital', 'represents nothing at all; he is a mere model of meekness' (pp.193f.). Towards the end of a long chapter on the limits and contradictions of Dickens' radicalism comes an appreciative note on *Crime and Punishment*, all the more suggestive when we relate it to Gissing's youthful law-breaking (Tindall, p.10) as well as to his fictional characters:

> Raskolnikoff..., a typical Russian, a man of brains maddened by hunger and by the sight of others hungry, is the kind of character Dickens never attempted to portray; his motives, his reasonings, could not be comprehended by an Englishman of the lower middle class. (*Charles Dickens*, p.214)

Many of Gissing's protagonists have 'brains'; so almost always do his narrators and implied authors. Taken together, these remarks in *Charles Dickens* remind us that for Gissing 'brains' connotes not just intelligence, but a critical sensibility, both socially determined and deliberately cultivated, that drives its possessor to reason about society and politics in ways that 'cannot be comprehended' by much of the public. 'Brains' and 'spirit of antagonism' are intimately allied. The best, most characteristic Gissing novels create the sense that language and ideas, far from being shared uncontentiously between writer and public, are contestable and disputed.

In illustration of this, we can consider the well-known description in *The Nether World* of the August Bank Holiday amusements at the Crystal Palace. Drawing a satirical but half-serious parallel with the Roman saturnalia ('Today will the slaves of industrialism don the *pileus*'), the narrator describes the beer-drinking crowd enjoying 'imbecile joviality'. There is a cocoa-nut shy where missiles are thrown at 'a wooden model of the treacherous Afghan or the base African' – 'the philosophic mind would have noted with interest how ingeniously these games were made to appeal to the patriotism of the throng' (*The Nether World*, pp.104, 108). A remedy is then offered:

> Well, as every man needs must have his panacea for the ills of society, let me inform you of mine. To humanise the multitude two things are necessary – two things of the simplest kind conceivable. In the first place, you must effect an entire change of economic conditions: a preliminary step of which every tyro will recognize the easiness; then you must bring to bear on the new order of things the constant influence of music. Does not the prescription recommend itself? It is jesting in earnest. (p.109)

We note the aggressive-defensive tone, the irony which makes issues more complex without fully clarifying the narrator's position, and indeed some uncertainty as to whether the narrator can be held finally to any position. The reader's agreement is not assumed; it is enforced, argumentatively. The argument is not straightforwardly either conservative or progressive. The plebeian holidaymakers are described at times with contempt, but the chapter places ultimate responsibility for their condition on large social forces: economic subordination ('slaves of industrialism'), and also the cultural manipulation to which Gissing was always very sensitive. These 'ills' can be remedied only by 'an entire change' whose plausibility, however, is left in doubt. The reader needs 'brains', then, to pick up the argument's nuances and engage with its dialogical address. Here there is some kinship with 'élite' writers (especially, perhaps, Meredith): Gissing's narrative and characterological patterns make few demands and bring few surprises, but his authorial discourse requires an alert response.

Gissing is prone to aggressive simplifications, but all across his work, as late as the response of Harvey Rolfe in *The Whirlpool* (1897) to Kipling's 'Barrack-Room Ballads', we find passages like the one just quoted, requiring us to think (*The Whirlpool*, pp.449f; see the discussions by David Glover and Simon James in the present volume). Gissing's most characteristic rhetorical address is not based on an assumption of shared values such as underlies the didactic wit of Austen, the authoritative judgement of Eliot, or Dickens' bravura passages of indignation, sentiment and comedy. Rather, we detect a resistant, embattled, insecure intelligence, actively making its case as we read, somewhat akin to that of Dostoevsky or Charlotte Bronte (whom Gissing much admired: see for example Gissing, 1929 [1898], p.193). Responding in 1893 to a questionnaire from *Pearson's Weekly*, he named his 'favourite novels' as *Villette*, *The Ordeal of Richard Feverel* and *The Return of the Native* (Gissing, 1994, p.169), all works in which the development of intelligence and character brings a rejection of conventional ideas. These novels, like several of Gissing's, reject the ideal-typical form and ideology of the *Bildungsroman*: growth leads not to integration but to alienation (see Moretti; Alden).

'Brains', which equip the thinker to think and the writer to write, also distance thinkers and writers from ready-made assumptions: so intellectual integrity tends to entail social marginality. Moreover, while the author's 'unclassed' status may be celebrated as a choice, it is also experienced as a fate. These are the decades during which the literary intelligentsia is coming to be, and to see itself as, 'the dominated fraction within the dominant class' (Bourdieu, p.317). Gissing has been a figure of recurrent interest partly because of his prescient concern with the ambiguities of intellectual and literary dissidence in conditions of 'mass culture'. Uncertainty about the author's claim to cultural authority underlies several of the unstable oppositions for which his work is noted: the 'negative identification' with the poor that flips over into vilification of the masses (the 'aristocracy of *brains*' versus 'the quarter-educated mob': Gissing, 1994, p.19), the vacillation between engagement with the world and retreat from it, the apprehension of writing and 'culture' both as commodity and as transcendence. Gissing's portrayal of women, which at times appears violently misogynistic but which is remarkable also for its interest in women's new social and personal opportunities, is one more ambiguity – and the most important site of his tendency to retreat, under pressure, to conservative positions that his own writing elsewhere exposes and undermines.

The problematic of the 'alienated intellectual' is fully and originally presented in Gissing's work; but it is presented more through contradictory implication and unresolved ambiguity than through clear and objective analysis. The 'man of letters' is shown, fittingly, as much in the texture of the writing as in any fixed and distanced portrait. Since for Fredric Jameson 'the aim of a properly structural interpretation or exegesis [is] the explosion of the seemingly unified text into a host of clashing and contradictory elements' (Jameson, pp.56f.), it is no wonder that he and other left-wing critics have found Gissing's novels amenable to our concerns, for here the 'clashing and contradictory elements' continually strain and burst through the text's seeming unity. Especially when he deals with questions of culture, gender and authority, the surface of Gissing's works often displays the anxiety that has driven him to write.

In the remainder of this chapter, I pursue the contrast between this figure of the writer as conflicted, ideologically engaged, and all too 'brainy' – the 'man of letters' in Gissing's fiction – and two other figures, whom I shall have to invoke rather schematically: the authoritative composer-ideologue of the high-realist 'unified text', and the 'portrait of the artist' as aesthetic technician. Calmly 'transmuting the daily bread of experience into the radiant body of everliving life' (Joyce, 1992 [1914], p.240), this artist-priest is the *miglior fabbro* (Eliot, 'Dedication'), the 'better craftsman' favoured in the self-image of literary modernism.

Writer-heroes

The voices of narrator and implied author in much of Gissing's fiction are energized, I have suggested, by 'brains' and a 'spirit of antagonism'. (Narrator and implied author are usually indistinguishable, and I shall henceforth speak simply of 'the narrator'.) A full account of the 'man of letters' in his work would have more to say about the construction of these voices. Moreover, many characters who are not writers – Godwin Peak, Rhoda Nunn, Harvey Rolfe (in *Born in Exile*, *The Odd Women* and *The Whirlpool*) – are intellectuals inasmuch as their sense of identity is created in a tensely reflective engagement with dominant cultural formations. I do not share Jameson's view that Gissing's 'greatest novels' are just those in which the problematic of the 'alienated intellectual' is specified in the figure of the writer (Jameson, p. 204). Still, my argument is most readily pursued in a discussion of writer-protagonists. I have already mentioned the four texts I shall consider: *The Unclassed* (1884), *New Grub Street* (1891), *In the Year of Jubilee* (1894) and *The Private Papers of Henry Ryecroft* (1903).

New Grub Street is of course a rich source of evidence about how Gissing conceived, and practised, writing as the expression of 'brains'. Christina Lupton and Tilman Reitz' essay in the present volume explores this in depth, so I say little about the novel here, though I shall mention Biffen and *Mr Bailey* later. I want simply to note that, surprising though this may seem, Gissing's most sustained account of literary production is given in a voice whose 'mature general tone' is striking and unusual (Williams, 1987, p.174). Here for once Gissing satisfies the formal and ideological demands of high realism. The complex perspective encompasses a range of socially-typical figures who convince us psychologically, as well as demarcating the novel's field of themes and values. *New Grub Street* strikes the reader as largely conforming with the ideal type of realist fiction as characterized by Elizabeth Ermarth:

> In realism, points of view must intersect to create a uniform horizon, disagreement must be resolved so that the final picture or narrative achieves consistency in all its relationships. Realism forces its readers and viewers into a 'middle distance' that makes possible the perception of a unified, single 'set of meanings'. (Ermarth, p.36)

In this 'seemingly unified text' (to recall Jameson's phrase), the author-narrator never obtrudes in immoderately challenging tones, but can imply or pass judgement with an ironic objectivity rare in Gissing. Given the novel's theme, this is surprising, I have suggested: but perhaps the equable tone and measured characterization were facilitated by the very directness and fullness with which the question of writing is confronted. (Analogously, *The Odd Women*, with its objectified character-images of a range of masculinities, femininities and partnerships, treats its central questions of marriage, feminism and sexual desire with greater calmness, less rage and anxiety, than we find in other Gissing novels where such questions are less fully addressed.)

Williams speculated that the 'mature tone' of *New Grub Street* marked 'an important stage in Gissing's development' (Williams, 1987, p.174). However, we find no such tone in *In the Year of Jubilee*, written three years later. David Glover, in his essay here, reads the latter novel in relation to wider currents in social history and literary representation, as an image of the metropolis of commodity and spectacle (themes central also to Glover's and Deborah Parson's readings of *The Whirlpool*). As it maps the encounter of a 'literary' sensibility with these new cultural conditions, the tone and perspective of *In the Year of Jubilee* (unlike those of *New Grub Street*) are marked by Gissing's characteristic instability.

This is reflected in the ambivalence of the writer-hero, Lionel Tarrant, towards the heroine, Nancy Lord (see Ryle and Soper, pp.127ff., 144ff.). The daughter of a Camberwell piano-dealer, Nancy becomes Tarrant's lover and the mother of his child. We are offered two ways of reading this relationship and judging the social and cultural questions it mediates. Tarrant with his Oxford education can be seen as the authentic masculine representative of literary culture, contrasted with his rival Luckworth Crewe the self-made advertising salesman. Charmed half against his will by Nancy, Tarrant must set about refining and disciplining her inauthentic, suburban, feminine notions of 'culture'. During their courtship, he pointedly gets her to read aloud from the book she has recently bought, Helmholtz's *Lectures on Scientific Subjects*, and then caps this by reciting Keats' 'La Belle Dame sans Merci' (*Jubilee*, pp.103f.). An opposite interpretation is also possible, however. In this, Tarrant's 'culture' is a mere adjunct of his privileged background – and is moreover compromised by the frankly commercial, journalistic use he decides to make of it (like Milvain or Whelpdale in *New Grub Street*). Against this, Nancy's secretly-written, painfully authentic and never published novel may be regarded as true literature. The narrator in some episodes seems to regard Nancy with sympathy, and to be intent on showing up Tarrant's hypocrisies and compromises. Elsewhere, however – and throughout the closing sequence – the narrative voice echoes and endorses Tarrant's need for control. The ironic and critical gaze that was earlier turned on Tarrant's cultural prestige falters, as the narrator identifies his own authority with that of the protagonist.

The novel's schema represents, in Nancy and other women of her milieu, much that Gissing found threatening: the growth of the suburban lower middle class (upwardly mobile, and acquiring a dominant position as cultural consumers despite what Gissing saw as their deficient education), and the destabilization of the patriarchal family as women enjoyed new sexual, educational and economic opportunities. The conclusion shows Tarrant having established with Nancy the marital arrangements he prefers: the partners will live apart, but he is free to visit her and their child just as he chooses. As Barbara Harman has argued (Harman, 1992), this settlement preserves his masculine autonomy, and her mistress-like desirability, while satisfying his need for absolute property in her. The novel colludes with Tarrant in its plot as well as its rhetorical emphasis, providing a solution that magically dispels anxiety: protagonist, and author, can appropriate Nancy and what she stands for on their own terms.

Critical emphasis should nonetheless fall on the unresolved, contradictory quality of the book. John Carey's hostile account in *The Intellectuals and the Masses* largely neglects this. Carey, reading *Jubilee* as a key document in his dossier of modernist and pre-modernist literary snobbery, can make his one-sided case with some plausibility because of the overt prejudices of the narrator, who ridicules the pretensions of those who claim new kinds of cultural literacy. For example, the description of the clerk Samuel Bennett Barmby as 'quite uneducated, in any legitimate sense of the word', vainly aspiring after 'Culture', which belongs to 'a world for ever closed against him' (*Jubilee*, p.180), is a crudely hostile response to the challenge of Barmby's milieu, with its mutual improvement societies, public meetings and discussion groups. Against this, the narrator anxiously defends his claim to represent 'legitimate', learned, literary culture. However, the treatment of petty-bourgeois life is not unremittingly hostile: Luckworth Crewe and the Brixton fashion entrepreneur Beatrice French are much more sympathetically presented than we might have anticipated.

What is then characteristic of this as of some other Gissing novels, but quite unlike any 'mature' perspectival realism, is the manifest presence of contradiction which has not been assimilated within a higher-order frame. The protagonist, focus of thematic concerns about class, culture, money, authority, cannot be held at stable distance. Because the narrator-author's anxieties are implicated in Tarrant's, the 'spirit of antagonism' prompts a narrative address that becomes direct, ideological, polemical – and inconsistent. These interventions exacerbate the impossibility of a unified reading, and confirm the unease they cannot assuage.

Such interventions, especially when they remind us of a too-close identity between novelist and character, disrupt the 'objective, impersonal language' proper to naturalism (Bowlby, pp.9f.). They infringe the norms of impersonality and indirection supposed to govern realist narration and which we find, canonically, in *Middlemarch* or *Anna Karenina*. In those texts, the novelist's ideological investments have been more effectively assimilated into the work's 'unity' and must be brought to light by the 'explosive' work of criticism.

If we ask what drives Gissing to represent (as he does in *Jubilee*) socio-cultural differences which then engender an ill-controlled vehemence, we may conclude that he is under some compulsion. So much is implied in *The Unclassed*, the first of his novels to be commercially published. (The production of *Workers in the Dawn* in 1880 was subsidized by the author: see Mattheisen et al, 'Chronology'.) The novel's schematic characterization counterposes poetical, Italianate Julian Casti, who aspires to write a Keatsian epic, and would-be social novelist Osmond Waymark. Casti, like his avatars Harold Biffen and Henry Ryecroft, is marked for death not life: as ineffectual as he is pure, he does not even complete his manuscript before dying of consumption. Waymark, as critics have noted, is an idealized self-portrait: it was under the pseudonym 'Osmond Waymark' that Gissing sent to Smith and Elder the manuscript of *A Life's Morning* (Mattheisen et al., 'Chronology').

In narrative terms, Gissing does not know what to do with Waymark the writer, beyond telling us that when his first book came out a few reviewers liked it, none 'heartily recommended' it, and several categorized him as an unsavoury naturalist (*The Unclassed*, p.290). *The Unclassed* ends with a fairy-tale resolution of Waymark's love-entanglements, giving little clue as to whether or how he will go on novel-writing. However, we can still find interest in his literary conversations with Casti, unsatisfactory though dialogue is as a means of developing thematic ideas. (Of course, the interest is partly contingent on the fact that Gissing was to become an important writer.) Raymond Williams, in a celebrated comment, discusses the speeches in which Waymark disavows his old radical commitments. These, he tells Casti, were the reflex of his own ' "vehement desires" ' and of a merely projective identification with the poor: ' "I did not make their cause my own, but my cause theirs." ' Williams notes that this illuminates the element of 'negative identification' in Gissing's work, and amounts to 'the best description' of that 'particular category of feeling' (*The Unclassed*, pp.211f.; Williams, 1987, pp.175f).

There is more to be said about this discussion. In a previous meeting, Waymark has told his friend: ' "Art, nowadays, must be the mouthpiece of misery, for misery is the keynote of modern life" ' (*The Unclassed*, p.165). (Gissing had quite recently described himself to his brother as 'a mouthpiece of the advanced Radical party', who would use fiction to expose social evils and governmental negligence: see Korg, p.28.) The later conversation comes in a chapter whose title, 'Art and Misery', announces that the unfinished business of the earlier dialogue will be resumed. Waymark now presents himself as liberated by a fuller self-understanding. He can put political anger behind him: ' "I have no longer a spark of social enthusiasm. Art is all I care for..." ' (*The Unclassed*, p.211).

Casti, however, has read Waymark's manuscript novel and been impressed by the ' "absolute truth" ' of its social depiction: ' "Such a book will do more good than half a dozen religious societies" ' (ibid.). He presses Waymark to examine more closely his claim to have found an entirely new vocation, as a writer rather than a 'ranting radical':

> "Yet," urged Julian, "the old direction still shows itself in your choice of subjects. Granting that this is pure art, it is a kind of art only possible to an age in which the social question is predominant."

> "True, very likely. Every strong individuality is more or less the expression if its age. This direction may be imposed on me; for all that, I understand why I pursue it." (p. 212)

'Individuality' and social determination are mutually implicated: what is 'imposed on' the self is constitutive of it, and this is true especially of the 'strong' individual. On the other hand, as was suggested in the phrase 'mouthpiece of misery', this self-formation may be felt as a constraint: as soon as the writer engages with 'modern life' and 'the social question', the self is open to what lies outside it, which may impose the 'direction' of the work.

As when he speaks of the writer for whom 'brains' are at once enabling and marginalizing, Gissing here seems close to articulating ideas that were later developed in the elaboration of more precise terms and concepts, specifically in Marxist cultural theory. The economic 'base', we might say, which Waymark cannot but register in its aspect of widespread 'misery', is seen to determine the cultural (or 'superstructural') work; but the latter nonetheless has its own autonomous standing. The dialogue, though it remains conversational and personal rather than theoretical, is interesting and suggestive. However, for his own mental equilibrium Waymark's resolution is unstable. 'The old direction' risks drawing him back not just to 'the social question' but to the 'enthusiasm' he claims to have outgrown. His boast – ' "I have no longer a spark of social enthusiasm. Art is all I now care for" ' – is suspiciously absolute: actually, Waymark-Gissing will go on imagining his engagement with the world to be transitive, effectual. 'Modern life' and 'the social question' connote literary subject-matter likely to provoke responses that will be critical and exhortatory rather than contemplatively mimetic. That Waymark's novel breathes this engaged spirit is confirmed by Casti's comment on it and by the published reviews with their references to powerful naturalism.

The discussion as a whole, in its full context, affirms what Waymark seeks to deny, namely the fundamental link between the artistic and the political sensibility. We must trust the tale, and its teller (the author-in-becoming George Gissing), rather than the wishful remarks of the writer-protagonist. A politically inflected sensibility will indeed speak through the characters, the narrator, and the schema of much of Gissing's best fiction. The implied politics tends to grow reactionary. Class difference becomes the prompt for anxious assertions of writerly superiority, and misogynistic sentiments and characterizations persist, belying any sense that in *The Odd Women* Gissing came to settled terms with feminism. However, the quality of Gissing's compound of 'social enthusiasm' with 'brains' is not essentially a matter of which opinions – progressive, or reactionary – are expressed or implied. The peculiar tone of his social fiction derives from the sense, expressed here, right at the start, that the enabling sensibility is also a troublesome burden. It is the recalcitrant, reluctantly-persistent voice speaking of 'the social question' that gives this unique tone, as of a realism ill at ease with its own terms (here is another aspect of the 'resentment' diagnosed by Jameson). Waymark says he is relieved to have got rid of 'social enthusiasm'. But in *The Private Papers of Henry Ryecroft*, it is still what the protagonist, and the author, say they are trying to escape from.

In this last work by Gissing to appear while he was alive, Ryecroft's diary is presented as posthumously edited, and the life it depicts is already 'a sort of death, or posthumous release' (Poole, p.205). The celebration of icons and sensations offered as 'rural England' comes with a show of literary conservatism, not to say dilettantism: 'Sitting in my garden amid the evening scent of roses, I have read through Walton's *Life of Hooker*' (*Ryecroft*, p.84). The day of writing is over. Gissing in his novels often enough invents characters and situations that body forth his preconceptions, but he also invents others to challenge and subvert them. In this formless pseudo-diary, however, Ryecroft encounters few fellow-beings, his docile housekeeper apart. The 'social question' is just the question of other people, far away and mostly unattractive. 'Social enthusiasm' nonetheless persists, as cantankerous sententiousness. The voice talks irritably of 'modern life': conscription, irremediable popular ignorance, the wrong-headedness of Republicanism, the adulteration of food – 'The deterioration of English butter is one of the worst signs of the moral state of our people' (pp.55f., 70f., 131, 249).

The notable point, beyond the unpleasantness of some of the opinions, is that even as he imagines a final release into rustic solitude and contemplation, Gissing drives his pen to write ideology and ideology-critique. Strip away all but the core of the writer's being, and what you find there will still be some trace, some version, of the antagonistic, brainy sensibility 'imposed' on the writer by the world.

Aesthetics and politics: Gissing, realism, modernism

If *Ryecroft* shows the persistence to the last of a resentfully politicized consciousness, it also offers that contrasting, ideal image of the literary sensibility enfranchised from the market-place. Freed by a legacy to do as he pleases, Ryecroft writes no more, and reads the classics – books English, German, Greek, Latin, almost all of them old. 'Literature' is the opposite of 'literary work' (pp.84f.), its purity guaranteed by its remoteness from contemporary social questions and its independence of the market.

This, if we took it seriously, would condemn the working novelist to practice a merely sub-literary trade. In fact, Gissing showed a finely discriminating sense of the possibilities open to the writer within the sphere of commercial fiction-publishing. However, there is little in his work – Biffen and *Mr Bailey* apart – to suggest that he saw the novel as a mode of fully literary art, or as susceptible of radical formal and aesthetic innovation. Classical authors and classical culture remain the touchstones to which he most often refers. His ironic reference to 'our grave Art of Fiction' shows, indeed, his awareness that there was 'a reappraisal of the possibilities and responsibilities of the novel' (Goode, 1966, pp.244f.) during his working life, which formed one context for pre-modernist experimentation. Rachel Bowlby points out that realist and naturalist writers who rejected romantic conceptions 'still needed to seek ways of attributing distinctive if not subversive power to art and artists' (Bowlby, p.95). However, Gissing's practice remains in formal terms workaday, conservative, a series of ad hoc adaptations of well-tried conventions. His letters of advice to his writer-brother Algernon and his other occasional comments on novel-writing offer little beyond sharp, particular critical observations and craftsmanlike technical hints (see Gissing, 1978). His work may pose interesting questions about how to read fiction, but we should not claim he was a deliberate formal innovator, or construe whatever is awkward in his novels as aesthetic virtue (James; Ryle, 2003).

The comparison with Hardy is illuminating. Both writers are in some ways highly original, and no other voices are like theirs; both audibly chafed at restrictive conventions and practices of novel-writing and -publishing; yet neither established distinctly new modes of narration and characterization. The habitual forms of realism may not quite fit their needs, but they work with them all the same, pragmatically and sometimes clumsily. A contrast can be drawn, however, in respect of their narrative voices. Hardy's narrators often strike us as less eloquent than the stories they tell. Authorial commentary, and modes of character-drawing and -analysis that depend on indirect discourse poised subtly between narrator and protagonist, are less well developed and less successful than story-telling and description. Hardy's great moments, whose emotional force is quite beyond Gissing's compass, very often draw on a visual and sensory realization that makes his scene not just the setting but also the 'objective correlative' of what takes place there. The young Jude's equivocal hilltop vision of Christminster, 'directly seen or miraged in the peculiar atmosphere', is one well-known instance of how this objective vision can create a complex of feelings and ideas complementing or replacing explicit commentary (Hardy, 1987, p.61). In Gissing, by contrast, the narrative voice is highly developed: over-developed, we might say, since the mediation of the narrator's 'strong individuality' seems often to bar the reader from direct vision, participation and identification.

However, this over-invested quality of narration in Gissing is intimately connected with his interest for later criticism. The self-consciousness that led him to 'show a man of letters' and that is audible in his voice is not merely personal. It reflects an awareness that the social basis of realist authority was being undermined (see Goode, 1998, pp.xviii ff.; Keating). 'Social questions', implicitly political, had been confidently addressed in the tradition. The formal organization of high realist fictions had been political in a deeper sense too, dependent on linkages between social hierarchies and ethical-epistemological perspectives, and between writers and audiences, as is agreed by critics whose valuations of realism are quite divergent (for example, Bersani, Ermarth, Lukács, MacCabe). Gissing, whose marginal position made him very sensitive to questions of authority, reflects in his work the developments that were weakening these linkages – diluting the ideological hegemony of 'the Middle Class, commonly called "Great" '; driving intellectuals to think in a 'spirit of antagonism'; and throwing writers into a marketplace where literary value might one day be fixed by a 'suburban' audience 'quite uneducated, in any legitimate sense of the word'.

Some of his novels review this situation explicitly, offering a perspective no other writer gives us. Its tensions also invest their tone and texture. This is because Gissing persistently returns, despite himself (it seems), to ideologically and politically inflected modes of writing. He cannot or will not relinquish the idea that novelists are to schematize, elucidate and comment authoritatively on 'social questions'; therefore he goes on writing realist novels. However, the very assertiveness of their narrators suggests that the authority of realist narration can no longer be equably sustained.

The most intriguing image Gissing gives us of another kind of writing is of course *Mr Bailey, Grocer*. Anyone thinking about the way forward from Gissing's kind of realism will reflect on *Mr Bailey* (as Patrick Parrinder does here). Gissing's description of the writing is scanty and unconvincing ('Each sentence was as good as he could make it', and so on: *New Grub Street*, p.426), but enough is said about Biffen's intentions and the book's reception to show us a kind of blank mould from which we can imagine it as cast. Of course, it will not offer the public what the public wants – the wifeless, childless Biffen can afford that 'kind of practical *bull*'. We must suppose it possesses great, though ill-defined, aesthetic distinction. Is this an anticipation, perhaps, of nineties decadence, *l'art pout l'art*, a variation on Gissing's other treatments of the theme of aestheticism (Maltz, 2001)?

The subject-matter of *Mr Bailey, Grocer*, petty-bourgeois and quotidian, hardly lends itself to decadent exoticism. In fact, the narrator refers to Biffen as 'the realist' – 'the realist would never smoke in Amy's presence' (*New Grub Street*, p.148). One phrase quoted from the few reviews that *Mr Bailey* gets might well apply to a novel by Gissing, or Waymark: it is 'another of those intolerable productions... indebted to the spirit of grovelling realism' (pp.485f.). However, in contrast to Gissing, Biffen (we are assured) writes at once realistically and dispassionately: with the dispassion, we might say, that Waymark hopes to have arrived at when he claims ' "Art is all I care for." ' Dedicated to ' "an absolute realism in the sphere of the ignobly decent" ', Biffen will avoid melodrama, indignation, humour, and any kind of commentary. He will treat

> "ordinary vulgar life with fidelity and seriousness... I am going to reproduce it verbatim, without one single impertinent suggestion of any point of view save that of honest reporting. The result will be something unutterably tedious. Precisely. That is the stamp of the ignobly decent life." (pp.144f.)

Here is one last set of interesting contradictions. Biffen/Gissing is surely too sophisticated (the reader objects) to think that 'verbatim honest reporting', without 'any point of view', can be written. Meanwhile the repeated phrase 'ignobly decent' in any case reveals strong negative preconceptions, aroused as ever by the spectacle of threatening cultural difference. We are, however, to imagine an improbable resolution of the antinomies of realism, which will erase authorial 'point of view' and authorial preconceptions, and preserve and enhance 'fidelity'.

Why then should the book, as Biffen describes it, not indeed be 'unutterably tedious' as he says it will be? After all, it is designed to repel the average reader and to make the average reviewer fulminate – ' "a pretentious book of the *genre ennuyant*... never interesting, never profitable... a novelist's first duty is to tell a story" ' (p.485). Here we see inscribed, but now in aesthetic terms, the writer's 'spirit of antagonism' towards his audience at large. Most readers will put the book aside. Those select few who appreciate it will have distinguished themselves by showing they can discern and value 'highly conscious artifice', of the kind that was to be seen as a key characteristic of the modernist artwork (Bradbury and McFarlane, p.24). Admiring art, they will not miss 'point of view', because in the coming fictional dispensation 'the content of the work of art, its "statement", recedes ever more as compared with the formal aspect, which defines itself as the aesthetic' (Bürger, pp.19f.). While we cannot imagine *Mr Bailey* as 'aesthetic' in the sense associated with the aestheticism of the 1890s, its 'formal aspect' is to predominate decisively over its 'content'.

John Goode comments that *New Grub Street* 'belongs as much to the modernist period as it does to the Victorian', and looks forward to the early work of Joyce and Woolf (Goode, 1998, p.xix). In one – negative – sense, this is true of much of Gissing's fiction. Persisting against the grain in making 'statements', and revealing the tensions this induces, he illustrates realism's constraints and norms more compellingly than writers who seem more accomplished. On these grounds, we can find his work more rewarding than most of Trollope's, for example. In it we feel realism at its limit, and sense the need for something to take its place.

In practice, Gissing nowhere tries even those limited innovations sketched in his account of Biffen's book. But if we read *Mr Bailey, Grocer* as an attempt to see beyond 'Victorian' novel-writing, a guess at the template for fictional modernism, we see that it not only misses the mark (as is hardly surprising), but hits it too. Gissing rightly anticipated that serious literary novelists would forge a new relation with a specialized public by making aesthetic and formal demands unwelcome to most readers. He was right, too, that overt opinions would less and less be voiced. The 'social enthusiasm' that had haunted the house of Victorian fiction would be much harder to detect in the new forms. The 'unified text' would be held airily aloft by aesthetic harmonies, rather than cobbled together from planks of opinion. Or so it would appear. Gissing reminds us, however, that the most triumphant innovations are from another point of view just the most adept adaptations to circumstance: 'This direction may be imposed on me; for all that, I understand why I pursue it.'

Bibliography

Alden, Patricia (1986), *Social Mobility in the English Bildungsroman: Gissing, Hardy, Bennett and Lawrence*, UMI Research Press, Ann Arbor and London.

Bersani, Leo (1976), *A Future for Astyanax: Character and Desire in Literature*, Marion Boyars, London.

Bourdieu, Pierre, trans. Richard Nice (1992), *Distinction: A Social Critique of the Judgement of Taste*, London, Routledge.

Bowlby, Rachel (1985), *Just Looking: Consumer Culture in Dreiser, Gissing and Zola*, Methuen, London and New York.

Bradbury, Malcolm and McFarlane, James (eds), 1978, *Modernism 1890 - 1930*, Harvester, Hassocks (Sussex).

Bürger, Peter, trans. Michael Shaw (1984), *Theory of the Avant Garde*, Manchester University Press and University of Minnesota Press, Manchester and Minneapolis.

Carey, John (1992), *The Intellectuals and the Masses: Pride and Prejudice among the Literary Intelligentsia, 1880 - 1939*, Faber and Faber, London.

Eliot, T.S. (1969; first publ. 1922), 'The Waste Land', in Eliot, *Complete Poems and Plays*, Faber and Faber, London.

Ermarth, Elizabeth Deeds (1983), *Realism and Consensus in the English Novel*, Princeton UP, Princeton.

Gissing, George (1929; first publ. 1898), *Charles Dickens: A Critical Study*, Blackie and Son, London and Glasgow.

Gissing, George (1994; first publ. 1894), *In the Year of Jubilee*, Everyman, London.

Gissing, George (1973; first publ. 1889), *The Nether World*, Everyman, London.

Gissing, George (1998; first publ. 1891), *New Grub Street*, Oxford U.P., Oxford.
Gissing, George (1903), *The Private Papers of Henry Ryecroft*, Constable, London.
Gissing, George (1911; first publ. 1884), *The Unclassed*, Sidgwick and Jackson, London.
Gissing, George (1984; first publ. 1897), *The Whirlpool*, Hogarth, London.
Goode, John (1966), 'The art of fiction: Walter Besant and Henry James', in David Howard, John Lucas and John Goode (eds.), *Tradition and tolerance in nineteenth-century fiction: critical essays on some English and American novels*, Routledge and Kegan Paul, London.
Goode, John (1998), 'Introduction', in Gissing (1998: 1891), pp.vii-xxi.
Hardy, Thomas (1987; first publ. 1896), *Jude the Obscure*, Penguin, Harmondsworth.
Harman, Barbara Leah (1992), 'Going Public: Female Emancipation in George Gissing's *In the Year of Jubilee*', orig. publ. in *Texas Studies in Language and Literature*, 34, extracts repr. in Gissing (1994: 1884), pp.389-393.
Huyssen, Andreas (1986), *After the Great Divide: Modernism, Mass Culture, Postmodernism*, Indiana University Press, Bloomington.
James, Simon J. (2001), 'Experiments in Realism: How to Read a Gissing Novel', in Bouwe Postmus (ed.), *A Garland for Gissing*, Rodopi, Amsterdam and New York, pp.11-22.
Jameson, Fredric (1983),*The Political Unconscious*, Methuen, London.
Joyce, James (1992; first publ. 1914-1915), *A Portrait of the Artist as a Young Man*, Penguin, Harmondsworth.
Keating, Peter (1991), *The Haunted Study: A Social History of the English Novel 1875-1914*, Fontana, London.
Korg, Jacob (1963), *George Gissing: A Critical Biography*, University of Washington Press, Washington.
Korg, Jacob and Korg, Cynthia (eds.) (1978), *George Gissing on Fiction*, *Enitharmon* Press, London.
Lukács, G., trans. Hannah and Stanley Mitchell (1969), *The Historical Novel*, Peregrine, Harmondsworth.
MacCabe, Colin (2002; first publ. 1979), *James Joyce and the Revolution of the Word*, Palgrave, Basingstoke.
Maltz, Diana (2001), 'George Gissing as Thwarted Aesthete', in Bouwe Postmus (ed), *A Garland for Gissing*, Rodopi, Amsterdam and New York, pp.203-213.
Mattheisen, Paul F., Young, Arthur C. and Coustillas, Pierre (eds.), (1994), *The Collected Letters of George Gissing: Volume Five, 1892 -1895*, Ohio University Press, Athens (Ohio).
Moretti, Franco (1987), *The Way of the World: The* Bildungsroman *in European Culture*, Verso, London.
Nicholls, Peter (1985), *Modernisms: a literary guide*, Macmillan, London.
Poole, Adrian (1975), *Gissing in Context*, Macmillan, London.
Ryle, Martin and Soper, Kate (2002), *To Relish the Sublime? Culture and Self-realization in Postmodern Times*, Verso, London.
Ryle, Martin (2003), Review of Postmus (ed.), *A Garland for Gissing*, *Nineteenth Century Literature*, 57, 4, pp.561-564.
Tindall, Gillian (1974), *The Born Exile: George Gissing*, Temple Smith, London.
White, Allon (1981), *The Uses of Obscurity: The Fiction of Early Modernism*, Routledge and Kegan Paul, London.
Williams, Raymond (1987; first publ. 1958), *Culture and Society*, Hogarth Press, London.
Williams, Raymond, ed. Tony Pinkney (1989), *The Politics of Modernism*, Verso, London.

Chapter 9

New Grub Street's Self-Consciousness

Christina Lupton and Tilman Reitz

The self-conscious character and the self-conscious novel

Early on in *New Grub Street*, the aspirant literary critic and journalist Jasper Milvain tells his future lover, Marian Yule, that he ' "shall do many a base thing in life, just to get money and reputation" ' (*New Grub Street*, p.119). Even at this stage of the narrative, the reader has ample cause to take this claim seriously. Marian, however, responds to Jasper's representation of himself sympathetically: 'She looked up at him with a smile. "People who are going to live unworthily don't declare it in this way." ' As it turns out, Jasper proves true to his word and does several 'base' things, including marrying Marian's cousin, Amy, whose inheritance guarantees him the material security he has openly pursued. But Marian's initial, sympathetic interpretation of Jasper's declaration remains pivotal in thinking about *New Grub Street* in the context of its own candour. By vividly detailing the commodification of literature, *New Grub Street* shares Marian's investment in the possibility that an open declaration of economic motivation functions as more than an act of self-indictment. Although Marian is wrong to imagine that Jasper transcends his self-interest simply because he can articulate this tendency himself, in the larger moral economy of *New Grub Street* she may be right to assert the redemptive value of expressing such knowledge. This paper wants to show that this holds true in two senses. Generally, we will suggest that the occurrence and articulation of knowledge about the way in which writing is produced changes the nature of this writing, namely the structure of the novel. But if this change should have any meaning for the way *New Grub Street* stands out in substance from Gissing's other fiction, the text must also work more precisely in the way that Jasper's speech act does for Marian: Gissing's describing the mechanisms of commercial literature should then also modify the attitude of his readers, address them as being less exposed to identification, and as more able to reflect on how literature might function in their lives.

John Goode reads Jasper Milvain's rationality as the unchallenged framework for Gissing's presentation of London's aspirant writers and critics, all of whom struggle to produce an increasingly market-oriented kind of literature. If Milvain doesn't stand out as a protagonist in the novel it is, he argues, 'because his rationality effectively becomes the novel's reality' (Goode, p. 117). In this reading, the difference closes up between Milvain, who writes articles in all too apparent awareness of the money to be made from them, and the failing novelist, Edwin Reardon, who loses his capacity to write fiction as soon as he must compete in the modern literary market: 'On the contrary... Reardon's whole story depends on his tacit acquiescence in Milvain's world view. He may lament the making of literature into a trade, but this is not because he has a literary integrity which offers an alternative view' (p.116). Like Milvain, Goode argues, Reardon sees literary production as a struggle in which the onus is on the writer to succeed. Both writers are portrayed as highly rational about the contexts in which they work; but while Milvain claims his rationality as a margin of freedom, Reardon takes an objectifying view of the struggle in which he is involved and resigns himself to rationally understanding his own failure in social Darwinist terms. Goode's argument makes the unusual move of stressing that Milvain and Reardon, who are usually seen as opposing kinds of writer, both support the idea that an increasing demand for witty reviews, snappy articles, and ' "specialism popularised" ' (*New Grub Street*, p.361) – pieces that can be read with an ever decreasing investment of time, attention, and education on the part of the reader – is inevitable. Although the two characters hold different views of art, in the end they occupy their positions in similar ways, exercising the possibility of astute, critical appraisal in relation to their own work while accepting the way literature comes to be valued and sold.

Both Reardon and Milvain possess an almost uncanny capability of assessing their own writing in terms of this market. Whether Milvain is predicting his own success as a ' "literary man of 1882" ' (p.8) or declaring with pride the efficacy of his day's work, its literary value ' "equal to the contents of a mouldy nut" '(p.181), his self-assessments are routinely confirmed as the public responds to his work. Reardon's view of his novels proves similarly prophetic. He wakes one night before his wedding in the fear, subsequently justified, that he 'could never get more than a hundred pounds for one of the long books which cost [him] so much labour' (p.66). After his marriage, Reardon writes his novel *Margaret Home* knowing that it is 'confoundedly bad' (p.129), accurately predicting that it will be read as weak in style and plot. Reardon's self-appraisal supports what Gissing's omniscient narrator tells us of the fictional work: 'Reardon's story was weak, and [his] second volume had to consist entirely of laborious padding' (p.131). Despite his increasing inability to write well, Reardon offers a stable perspective from which he registers his changing abilities as a writer and confirms that the price given for his work reflects its quality.

Milvain and Reardon so frequently express verdicts about their own work and actions that are in agreement with the perspective of the narrator that it is easy to forget the special property of self-consciousness Gissing has granted his characters. All the writers in *New Grub Street* are, in fact, endowed with unusually realistic degrees of knowledge about their own chances of success. Alfred Yule, who is briefly deluded about his ability to succeed as the editor of a new journal, is able to describe his own faults as a writer of fiction: ' "I never quite knew what I wished to be at in fiction. I wasn't content to write a mere narrative of the exciting kind, yet I couldn't hit upon subjects of intellectual cast that altogether satisfied me" ' (p.398). And Whelpdale, who writes a successful manual of literary advice, confesses openly to his future wife, Dora Milvain, that he lacks the talent of her brother, Jasper: ' " I point the way," ' he says, ' "but I haven't been able to travel it myself" ' (p.273). In general, although Gissing's characters often describe their places within the social and economic hierarchy as unfair, they also internalize the language of modern literary publishers and the reading public. Their expressions of resentment against the system in which they live are undermined by the sense of objectivity that they grant this system when confirming its aesthetic judgements: if Reardon is angry that he is poor, his anger is weakened by the fact that he feels his own writing to be increasingly undeserving of recognition. With the possible exception of characters like Milvain, whose acute perceptions of society help them to actively align themselves with its prevailing conditions, Gissing therefore resists representing consciousness as something that helps to transform the selves or situations people become conscious of. On the contrary, *New Grub Street* cultivates a disconnection between what it means to observe oneself being exploited as a producer, and what it means to act against the injustice of this. This disjuncture is fundamental to *New Grub Street*'s narrative. Through characters who represent their own work and their economic positions to themselves, Gissing illustrates that grasping a situation intellectually does not prevent one from being determined by it.

Exposing the labour of writing

The predicament described here pertains to the characters in *New Grub Street* – although they know their own situations well, this knowledge has no function to play in improving their highly determined lives. But the same predicament can be seen as central to the form of Gissing's novel. John Peck has intimated, not very enthusiastically, the possibility of approaching *New Grub Street*'s reflexivity formally:

> [T]here is one aspect of *New Grub Street* which offers a tempting promise for formalist critics, attracted as they are by novels in the *Tristram Shandy* tradition, that is, novels about novel-writing, and this is the possibility that the novel itself, in its form, is the perfect illustration of the problems about novel-writing raised in the context of the novel. That is to say, we understand Reardon through the experience of reading a work in which all the strains of writing a three-volume novel are in evidence. (Peck, p.147)

The limited parallel Peck points to hinges on the mirroring of the real circumstances under which Gissing wrote *New Grub Street* in the storyline he developed for Reardon. Detailed descriptions of Reardon's increasingly desperate daily practice, his poised pen, and his erased sentences, open up what Benjamin Lee has described as the performative, 'metalinguistic "space" introduced by speech about speech and discourse about discourse' (Lee, p.11). In his diary, Gissing records writing *New Grub Street* in 1890 between October 6th and December 6th, a short eight weeks which testify to his training in efficiently turning out the three-decker novels by which he earned his often subsistence-level living. This timeframe corresponds almost exactly with the period in which *Margaret Home* is written – a novel Reardon begins at the end of the summer of 1882 and, after several false starts (analogous to those Gissing made before writing *New Grub Street*), finishes well before Christmas. Reardon, who can write four slips a day, with fifteen days for a volume and forty-five for a completed three-volume novel (*New Grub Street*, p.121), thus becomes a cipher for the technique Gissing himself developed as a writer.

Spelling this doubling out, however, does not describe its significance for the kind of novel *New Grub Street* becomes. Peck's assertion that it places *New Grub Street* in the '*Tristram Shandy* tradition' is only of limited promise. The potential comparison is offset not only by the clear difference in mood that divides Sterne's playfulness from Gissing's despondent record of a writer's life, but also by the technical differences in the two projects. In the mid-eighteenth century, when the conventions of novel writing, autobiography and history writing were not yet established, Sterne wrote about a character failing to write his autobiography. Through its self-consciousness about the process of writing an autobiography *Tristram Shandy* exposes the contingencies of personhood inherent in the emergent claims of empiricism, and of novel writing, while illustrating at the same time that irrational, subjective life can be communicated and enjoyed. Although Sterne frequently makes the metalinguistic move of producing discourse about discourse, or referring to the existence of his reading public and to his critics as practitioners in the process of his novel's getting made, what he exposes through these references is his place in a sensible community of readers, rather than an estrangement between readers, or his own estrangement from his published text.

In contrast, *New Grub Street* is self-conscious about precisely this process of estrangement. Gissing describes a range of fiction writers, critics, and writers of scholarly articles, united in their experience of the way that the capitalist market transforms art and its critique into a commodity. The publishers of Reardon's novels and articles, as of Jasper Milvain's witty social and literary journalism, Alfred Yule's prosaically researched pieces, and Dora Milvain's women's fiction and children's history, remain largely anonymous purchasers of words. As literary producers, these characters must anticipate the needs and desires of the market – at worst they must cater to the ' "quarter-educated" ', the readers of Whelpdale's journal *Chit-Chat*, who will ' "soon be abundantly provided with literature to their taste" ' (*New Grub Street*, p.478) and at best to readers like Amy Reardon, who dismisses fiction altogether in favour of journal articles. As Reardon sits at his desk composing his three-volume novel, he imagines, not the community of interests, values, and knowledge that unites the writers around him, but the depersonalized world of the market, of which he has only the lowest expectations, and on which he is totally reliant for his livelihood. *New Grub Street* describes the poorest producers of art; those most alienated from the commodified form literature takes. Their lack of income, leisure time, and access to the positions of power (for instance, to the editorship of one of the prominent journals that determine literary taste) means that Reardon, Biffen, Yule, and even the more successful Whelpdale and Milvain, experience the pressure of the market as a direct restraint. Although Reardon idealizes the possibility of artistic creation unfettered by need, describing the time when he was in fuller possession of his own capacities as a writer because he had less investment in the output of his labour, it is finally only Biffen who writes exactly as he wishes – an approach that his suicide underscores as literally unsustainable. Not only do these writers have very limited access as consumers to the products writers produce, they also have very little control over how they spend their energy as artists.

At the point where Reardon confronts his proof for *Margaret Home* he decries his alienation from his own text, measuring the distance the manuscript has travelled from him by imaging briefly the position of its consumer: ' "What hellish torment it was to write that page! I did it one morning when the fog was so thick that I had to light the lamp. It brings cold sweat to my forehead to read the words. And to think that people will skim over it without a suspicion of what it cost the writer!" ' (p.201). Reardon's anger arises less from a defence of the novel, which he knows is weak, than from the fact that it will appear as a commodity to its reader – an object whose identity hinges on its concealing the labour its existence has cost. When he reads a critique of *Margaret Home* claiming 'it contained not a single striking scene and not one living character', he agrees with his critic and admits the assessment could have been his own (p.207). But, again, he rails against the distance upon which such objective critique is premised: he recognizes that 'the critic of an isolated book has of course nothing to do with its author's state of mind and body any more than with the condition of his purse', but points, nevertheless, to the truth the critic is structurally obliged to overlook: 'not he had written this book, but his accursed poverty' (p.207). The distance of the critic, who views the work of art as a given object, becomes analogous to the distance of the rich who, for Reardon, simply see differently to the way he sees. Wealth endows people with the enviable ability to screen out considerations of labour when they observe objects and people around them (p.199).

In this context, the particular role of *New Grub Street*'s self-consciousness is that it challenges the ease with which the modern reader approaches the novel as an object fixed in value by the impersonal forces of the market. Print capitalism is not entirely responsible for the alienation of works of literature from the reality of their production and reception – the earlier idea of divine inspiration is, for instance, analogously hostile to recognizing the process of human creation. But under the conditions of print capitalism, the objectification of works of art conspires with the idea of aesthetic judgement as a response only to the work itself. Gissing works against this conspiracy when he makes the description of writing *New Grub Street* part of its final form. In this sense, he does something parallel to what Marx set out to do in the first volume of Capital: he reveals the labour embodied in the commodity *New Grub Street* becomes. By writing about writing, Gissing protests the possibility that his novel will be 'skimmed over' as something objectively given.

At the same time, in producing *New Grub Street* for the market he describes, and in which it sold better than his previous novels, Gissing embarks on this critique within the limits he observes, and does not deny that his own novel must also appeal to the market. This kind of exposure shares the gesture of 'honest dishonesty' associated more generally with the self-conscious novel. But what novelists, from Cervantes to Borges, have been conscious about is the limits of representation – the lie inherent in the artifice of fiction. In Brian Stonehill's definition, self-conscious novels 'acknowledge what they are' by making readers privy to the process of writing, forcing them to recognize the fact that novels are not real: 'there is thus an alienation of the reader from the novel's action at one level – these are, the reader sees, characters and actions invented by the novelist solely for purposes of his or her own – while at another level the reader, by being made conscious of his or her role as a listener confronted by a storyteller, is drawn into a stronger bond of intimacy' (Stonehill, p.7). This process of alienation and reconciliation occurs as the tenuous relation of language to reality is acknowledged. Open concessions to the material forces driving fictional representation are, however, relatively rare in this literary tradition. Asking what self-consciousness does to the structure of *New Grub Street* therefore brings together two questions that have until now been raised on distinct fronts: one, the question of what self-consciousness within fictional representation does to the genre of the novel, and the other, the Marxian question of the effects on human relations of a consciousness of commodification.

Changing society through self reflection

This is not to say that critics have not approached the role self-consciousness plays in Gissing's novels more generally. John Goode and Fredric Jameson have both described Gissing registering the loss of the totality that was historically offered by the perspective of the realist novel. Even if Gissing's protagonists are extremely conscious of the social spaces and operations around them, this understanding no longer promises material triumph for the hero, nor a formal one for the author. In Goode's terms, Gissing can be placed alongside Flaubert and Dostoevsky, writers who grapple with the realization that conditions of production can become clearer and more determining at the same time:

> Contiguity and totality no longer articulate themselves in ideology as "consciousness" as soon as productive labour and marriage, the forms of bourgeois oppression, constrict the protagonist to a place which is more rigid than the relatively sartorial place offered by a disintegrating feudal world. It is no longer possible to postulate self-discovery as totality when what you have to discover is that you are Madame Bovary. (Goode, p.38)

Goode sums up his account of *New Grub Street* by pointing out that the novel exposes the inability of the literary market to survive without the use of legalized expropriation. The function of Gissing's exposure, he adds, 'is not to oppose this "truth" but to reveal its way of working' (p.141). Jameson, too, describes Gissing representing an increasingly determined universe while developing a new strain of consciousness – in particular, the consciousness of the intellectual alienated from his class – that registers but does not exempt itself from these constraints. In his later novels, Gissing develops a style, even a wit, which transforms what Jameson terms his characteristic, early '*ressentiment*' of the upper classes, whose attitudes he both identifies with and testifies against. *Ressentiment* reaches its limits in Gissing's later novels, Jameson suggests, as an 'Unhappy Consciousness' which captures this state of ambivalence from an omnipresent but unproductive vantage point. Jameson's account of Gissing's work thus suggests the disintegration of consciousness as a viable engine of contradiction or change (Jameson, pp.203f.).

In the background of these accounts of what consciousness doesn't do in Gissing's fiction is not only the paradigm of earlier realist narratives, but also the impact of nineteenth- and twentieth-century philosophical accounts of what consciousness might do for history. As concepts like 'unhappy consciousness' and 'totality' indicate, both Jameson and Goode are working with a Hegelian-Marxist account of history as it has been shaped by Georg Lukács and early Critical Theory. In following this tradition, Jameson and Goode both appeal to a more rigorous and less pessimistic view of self-reflection than their analyses of Gissing suggest. Indeed, the description of consciousness as a transformative force not only drives the Hegelian tradition of dialectical thought; it goes to the heart of what Hegel's dialectic offers to the Marxian tradition. For Hegel, self-consciousness structures the reality it seems at first only to refer to: 'Consciousness appears split into this given actuality and the End which it realizes by superseding that actuality, an End which, in fact, makes an actuality in place of that which was *given*' (Hegel, p.216). In this way, every new insight into the situation of the conscious self or – later – the community of consciously acting subjects alters the situation: this is the very process which keeps the *Phenomenology of Spirit* going. Hegel's insight into the transformative role of consciousness had direct import for the notion of a worker's consciousness of him or herself as a commodity. Lukács closely combines the materialist and the idealist vocabulary: 'By adding self-consciousness to the commodity structure a new element is introduced, one that is different in principle and in quality from what is normally described as consciousness "of" an object' (Lukács, p.168). This is, Lukács underscores, because '*this knowledge brings about an objective structural change in the object of knowledge*' (p.169); the proletariat's act of establishing a 'self-consciousness of the commodity' is the first step in abolishing the commodified society. In contrast with how the process of history is represented in *New Grub Street*, Lukács gives an account of change in which the worker's becoming conscious cannot be distinguished from the process of his or her liberation: consciousness is not a precursor to change but its very essence.

To be sure, the absence of such logic in *New Grub Street* is not simply Gissing's shortcoming. The application of Hegelian-Marxist dialectics to the novel proves difficult for two reasons. Firstly, the step from self-consciousness in general to politically decisive knowledge is also difficult in Lukács's construction: while Hegel can easily treat transformations in the structure of knowledge as real changes because he is asking about the history or 'experience' of (self)-consciousness (Hegel, p.12), Lukács has to substantiate the Marxian dictum that 'theory itself becomes a material force when it has seized the masses' (Marx, 1978, p.60). The questionable point in his account is that Lukács obviously sees the possible epistemic superiority of the proletariat as a main factor in its imminent political victory. In this sense, *New Grub Street*'s view of society simply proves more realistic – one is tempted to say: more materialist – than the theory of revolution through self-consciousness. But secondly, if one reduces the hubris of such revolutionary logic to the simple but plausible argument that a group will only be able to act when its members recognize their shared interests, it seems quite clear that the commercial novel's self-reflection is not the most likely place for revolutionary self-consciousness to develop. Politically effective knowledge is to be expected mainly from political organizations – for Lukács and many of his contemporaries: the Communist Party – in contrast to which the self-reflection of writers and texts only accentuates and deepens the commodified state of affairs.

While the pre-Marxist Lukács of the *Theory of the Novel* already saw that the lone protagonist of modern epic fiction had the 'orientation (*Gesinnung*) towards totality' (Lukács, 2000, pp.145f.) but continually failed to reach this point, his later political theory, where he claims to have found 'a vantage point from which to survey the whole of society' (Lukács, 1968, p.20), explicitly situates commercial writing on the other, fragmented side:

> The specialized "virtuoso", the vendor of his objectified and reified faculties does not just becomes the [passive] observer of society; he also lapses into a contemplative attitude vis-à-vis the working of his own objectified and reified faculties.... This phenomenon can be seen at its most grotesque in journalism. Here it is subjectivity itself, knowledge, temperament and powers of expression that are reduced to an abstract mechanism.... The journalist's "lack of convictions", the prostitution of his experiences and beliefs is comprehensible only as the apogee of capitalist reification (p.100).

Tracing reification's benefits

There is, however, a different strain of Marxist theory which is fascinated exactly by lost social characters of this kind, floating between over-adaptation and the acute knowledge of their own exploitation in positions quite similar to ones represented and performed in Gissing's fiction. While some of its features have been spelled out by Adorno, the author to instigate this strain of theory is Walter Benjamin. In a development strikingly similar to (and not without being influenced by) the work of Lukács, the later Benjamin re-stages problems first formulated in literary theory as complexes of history and class consciousness – but denies the prospect of a solution. The theme of self-consciousness remains present in Benjamin's work from the point where it appears in his dissertation about the Early Romantic concept of art criticism, which he deciphers as the idea that the critique of a literary text is only a continuation of this text's self-reflection. Yet class consciousness comes in from another corner; it builds on his peculiar combination of hope and absolute despair. Although he was not yet acquainted with the theory of reification when he wrote his *Origin of German Tragic Drama*, it is here Benjamin first describes a situation where human relations are treated like things, where things have become utterly manipulable, and where any perspective of meaning in history therefore has vanished. In the world of baroque allegory, there is a 'primacy of the thing over the personal, the fragment over the total' (Benjamin, 1977, p.187); it presents 'everything about history that, from the very beginning, has been untimely, sorrowful, unsuccessful' (p.166). If Benjamin is fascinated by such a vision from early on, and later transforms it into a theory of nineteenth-century commodification processes, this is not simply because it appeals to his experience of his own time (which it clearly does). It corresponds, moreover, to his strategy of making the inevitability of hopelessness evident for the sake of breaking its spell. In order to really change things, so his famous late reflection goes, one has to recognize that the 'the "state of emergency" in which we live is not the exception but the rule' (Benjamin, 1968, p.259). This is the general structure underlying Benjamin's specific casting of the self-consciousness of the commodity: the prostitute, who figures as its 'humanization (*Menschwerdung*)' (Benjamin, 1972, vol.1, p.671); the flâneur, whose experience is shaped by the aesthetics of advertising and shop windows and thus by 'empathy towards the commodity' (1999, M 17 a, p.2); and finally the commercial writer who offers his reified experiences and convictions on the literary market. Benjamin even places Baudelaire in this functional context: being the first poet to follow the 'conception of a marketable originality' (ibid., J 58, p.4), he 'threw himself after his work and thus made true for his person... what he thought about the inevitability of prostitution for poets' (ibid., J 60 a, p.3).

This roughly explains Benjamin's sympathetic view of the position Gissing occupies as an author. At the same time, though, the liberating perspectives present in Lukács seem to have vanished: the idea that the social order will be changed as soon as its inhuman character has become thoroughly visible now seems hardly more than wishful thinking. Benjamin's work, however, offers two possibilities for giving a more concrete shape to the emancipatory potential of reification once it is realized. The first of them is widely known, but of comparatively little help: starting with his early essay on Goethe's *Wahlverwandtschaften* and continuing through to his 'Theses on the Philosophy of History', Benjamin unfolds a model of redemption coming from outside; from a posterity recognizing and rescuing the ones who had no chance to free themselves or were defeated in their struggle for liberation. In this context, the analysis of reification that occurred in the past coincides with realizing that there is something to be redeemed, and that this redemption is linked to one's own struggles and hopes. Yet this conception is not only based upon a quasi-theological belief in some Messianic end of history; it also separates those suffering a claustrophobic situation from those capable of changing it by reflection. Benjamin's account of the liberating reflection differs from Hegelian and Marxian models of transformative self-consciousness because, in his account, it is not the original agents who transcend their own situation. This takes us back to Benjamin's old idea that the reflexive process is not restricted to a text or, now, to a group of agents, but expands as long as there are interpreters maintaining a relation towards it. In order to see *New Grub Street* as emancipative in these terms, one would have to locate its consciousness as the property neither of Gissing nor of his characters, but of future readers precluded from the very process of reification it represents and enacts. Such a strain of interpretation fits quite well with the insistence of many critics that Gissing exposes the contradictions of late-Victorian culture without being fully in control of his own exposé (c.f. Harsh, p.873; Jameson; Selig, p.720). But this is an interpretation that *New Grub Street* seems very precisely to resist by announcing its capacity to itself reflect on its own conditions of production – an assertion of control that defies distinctions such as the one made by Robert Selig, whose reading of *In The Year of Jubilee* ends by separating the mistakes of 'Gissing the conscious thinker' from the intuitive awareness of 'Gissing the novelist' (Selig, p.720).

Benjamin's intimations about the positive potential of the commodified world do not preclude, however, the possibility that agents may act on their own behalf. Contrary to the line taken by Lukács and Adorno, which demonstrates the repressive character of enlightenment itself, Benjamin also tries to show what can be liberating about a determined process of rationalization. The central features of this dialectic all apply to the case of the market-oriented writer. Firstly, the reification of the human and the de-qualification of the object world not only bring about an impoverishment of life, but also destroy ideological taboos. Under these conditions, the work of art is no longer an untouchable, quasi-cult object, but may be used and criticized by everyone; the free individual, the hub of all liberal ideology, may be seen as fulfilling different social functions and being divisible into heterogeneous elements. At several points – namely in his essays 'The Work of Art in the Age of Mechanical Reproduction' and 'The Author as Producer' – Benjamin predicted that the new accessibility of the work of art would become socially useful once the roles of the art producer, the critic, and the audience were no longer separate. Gissing's contradictory work as producer and critic lays out what this heterogeneity of roles might look like. In this light, the metadiscursive level of *New Grub Street* – its discourse about discourse – allows Gissing to perform the roles he represents, of writer, critic, and producer, while maintaining an identity distinct from any one of them. The fact that these roles are divided out problematically, and somewhat arbitrarily, within the novel suggests that Gissing was motivated to explore an alternative, more coherent division of labour. Within the novel, Reardon, who had initially set out to write non-fiction, becomes a novelist only at the suggestion of a writer who does not know him at all. Despite his evident ability to critically review his own work he must subsequently rely on critics like Milvain for public judgement. Marian Yule, who works doggedly as her father's research assistant, becomes the physically exhausted producer of literary texts but is denied the role of author. In fictionally representing his own writing process, Gissing actively puts these artificially separated roles into dialogue by creating a structure in which an author who is forced to work as a producer may simultaneously occupy the role of his or her own critic.

Secondly, for Benjamin, the alienated structure of time which is forced upon the industrial as well as on the cultural-industrial worker not only splits up experience into a series of shock moments, but also makes for the possibility of changing human habits more radically and more quickly than ever before. Benjamin discusses this dialectic not only in the case of the film but again also in that of Baudelaire's poetry, where the permanent production of shock moments results from and calls for an ever-alert 'consciousness' or 'presence of mind' (1968, pp.162ff., 240). It is a similar sort of acuteness that characterizes the judgements of Milvain, Reardon, and Whelpdale. Their need to calculate the marketability of their own writing in an accelerating society produces reflections that might be described as sociological perceptions. Having announced that his paper, *Chit-Chat*, should address the ' "quarter-educated" ', Whelpdale goes on to explain: ' "That is to say, the great new generation that is just being turned out by the board schools, the young men and women who can just read, but are incapable of sustained attention. Everything must be very short, two inches at utmost; their attention can't sustain itself beyond two inches" ' (*New Grub Street*, p.496). However cynical the description may sound, it announces a new awareness of people's habits being shaped by changeable social settings, and it suggests how completely these habits have been stripped of any traditional fixation.

Finally, the writer's and producer's objectifying depiction of the reading audience also proves potentially liberating in its very cynicism. To be sure,

Whelpdale intends to address less a structured community of responsible subjects than an anonymous and de-centred mass. But this mass doesn't have to be a mere object of domination. Benjamin's famous title for the production of the capitalist mass audience is 'distraction' (*Zerstreuung*), which implies both the function of entertainment and the social structure of disorganization. Benjamin wants them both, and not only because they seem to be the last resort against fascist domination. In hailing the distraction of the audience, he aims at a quasi-normalization of Brechtian estrangement – a process that works against emotional identification for the sake of self-distancing understanding. The conscious readers of Gissing's self-conscious novel have to realize that they are consuming a mass product manufactured to attend to their particular needs in only the most calculating terms. Such readers find themselves described in the professional language of Milvain and Whelpdale as predictable customers, with every aspect of their desire for distinction calculable. In this way, the positive possibility of identification is subverted and any pre-given sense of identity is dissolved. Although this negative work of presenting readers to themselves in unfamiliar terms does not engage Gissing in political organization, the moments of self-consciousness he produces are the prerequisite to any new, revolutionary formation taking place.

In describing what *New Grub Street*'s consciousness of itself as a commodity does to its role in the situation it represents, we are left, then, with the question of what effect its structure has on readers. Benjamin's intimations of the radical potential of mass society help to reorient this line of questioning: we are no longer asking (with Marian Yule) the idealist question of whether consciousness transcends the totality it comprehends. Of course, to the extent that the nature and function of the commodity is defined by its concealment of its own function, any commodity that announces this function will cease to be a commodity in quite the ordinary sense. But as the commercial success of *New Grub Street* suggests, there are many cases where degrees of self-consciousness are compatible with the expectations of those consuming the literary commodity, and certain aspects of social critique sold as well on the late-Victorian culture market as they do today. It is perfectly possible, in other words, that we take Gissing at his word about the fact that the novel and the writer have become commodities, and that they stay commodities, perhaps even fulfil the requirements of art as a commodity in certain ways, as they identify their own nature. In this sense it is difficult and perhaps pointless to insist that the articulation of 'base' truths speaks automatically to the transformation of these truths.

Once this is granted, however, readers' response to a novel announcing its own commodification remains a distinctly possible field of transformation. Perhaps precisely because, as Goode points out, *New Grub Street* explicates its status without any pretension to changing it, it transfers the task of change to its audience; it points to the practice of reading as something that makes us culpable. The most immediate effect that the novel has in this respect is that it undoes the myth of the objectively given work of art (which is linked historically to the myth of the objectively given commodity) by making readers conscious of the ways in which realist narratives are made. This does not, however, produce the climate of intimacy that Stonehill suggests is generated by self-conscious novels generally.

New Grub Street's exposure of its own conditions of production is more an uncomfortable accusation of involvement than a cosy invitation to join the writer behind the scenes of his own fiction; one that also intimates that sympathy may not count for much in the material practice of culture. Such discomfort has its part to play, though, in the positive phenomena Benjamin attributes to consumer society – both in the breaking of the taboos surrounding art and in the shock effects of self-estrangement. Thus, while self-consciousness as such, of characters or of commodities themselves, offers no guarantee of change in the world Gissing represents, *New Grub Street*'s very particular structure of self-consciousness inscribes a place for the reader's increasing consciousness about the novel.

Bibliography

Benjamin, Walter, trans. John Osborne (1977), *The Origin of German Tragic Drama*, New Left Books, London.

Benjamin, Walter, ed. R. Tiedemann (1972), *Gesammelte Schriften*, Suhrkamp, Frankfurt.

Benjamin, Walter, trans. Harry Zohn, ed. Hannah Arendt (1968), *Illuminations*, Shocken Books, New York.

Benjamin, Walter, trans. Howard Eiland and Kevin McLaughlin (1999), *The Arcades Project*, Belknap Press, Cambridge (Ma).

Gissing, George (1993; first publ. 1891), *New Grub Street*, Oxford University Press, Oxford.

Goode, John (1978), *George Gissing: Ideology and Fiction*, Vision Press, London.

Harsh, Constance D. (1994). 'Gissing's *In the Year of Jubilee* and the Epistemology of Resistance', *Studies in English Literature, 1500-1900*, Volume 34, Issue 4, pp.853-857.

Hegel, G.W.F., trans. A.V. Miller (1977), *Phenomenology of Spirit*, OUP, Oxford.

Jameson, Fredric (1996), *The Political Unconscious*, Routledge, London.

Lee, Benjamin (1997), *Talking Heads: Language, Metalanguage and the Semiotics of Subjectivity*, Duke University Press, Durham and London.

Lukács, Georg, trans. Rodney Livingstone (1968), *History and Class Consciousness: Studies in Marxist Dialectics*, The MIT Press, Cambridge (Ma).

Lukács, Georg, ed. Michael McKeon (2000), *Theory of the Novel: A Historical Approach*, Johns Hopkins University Press, Baltimore.

Marx, Karl and Engels, Friedrich, ed. Robert C. Tucker (1978), *The Marx-Engels Reader*, Norton, New York.

Peck, John (1981), '*New Grub Street*: An Approach through Form', in Jean-Pierre Micheaux (ed.), *George Gissing: Critical Essays*, Vision Press, London.

Selig, Robert (1969), 'A Sad Heart at the Late-Victorian Culture Market: George Gissing's *In the Year of Jubilee*', in *Studies in English Literature, 1500-1900*, Volume 9, Issue 4, pp.703-720.

Stonehill, Brian (1988), *The Self Conscious Novel: Artifice in Fiction from Joyce to Pynchon*, University of Pennsylvania Press, Philadelphia.

Chapter 10

The Voice of the Unclassed: Gissing and Twentieth-Century English Fiction

Patrick Parrinder

Gissing as precursor

'Poor Gissing struck me as quite particularly marked out for what is called in his and my profession an unhappy ending' (Allen [attrib.], p.43). Henry James's reported remark suggests that Gissing's contemporaries must have thought of him as an Awful Warning rather than as some kind of role-model. Soon after his death, close friends such as H.G. Wells and Morley Roberts propagated the legend that Gissing should never have written novels at all. He seems, then, an unlikely mentor for successive generations of novelists in the twentieth century. He was neither a proto-Modernist, nor a writer of social comedy, nor a practitioner of the Jamesian Art of Fiction. His books lacked wit and romance, according to Virginia Woolf in 1912, however much they might offer 'thought and understanding of life as it is' (Woolf, 1972, p.530).

But Woolf wrote at least four essays and reviews on Gissing; one of these, her 1927 review of the Gissing family letters, has been much reprinted. Other novelists who have written memorably about him include Arnold Bennett, George Orwell, William Plomer, V.S. Pritchett and Gillian Tindall. Not all these writers paid him the compliment of imitation, though some did. Among Gissing's critics it was Orwell (whose credentials as a novelist have been questioned nearly as often as Gissing's) who most clearly looked to him as a precursor. When Pritchett described Gissing as the 'perverse prophet and progenitor' of the new English novelists of the 1950s (Pritchett, p.63), it was, perhaps, Gissing's presence behind Orwell's direct influence that he was measuring. And this presence, as the century went on, became a penumbra rather than a shadow. Yet, as we shall see, Gissing's personal example, his presentation of the unclassed intellectual, and his realist aesthetic were not forgotten. At the same time, what had generated a tragic attitude in Gissing's best-known novels became a source of comedy and outright farce; and Gissing's contempt for the vulgarity of ordinary suburban life became transmuted into affection and even nostalgia.

George Gissing appears as a named precursor in a first novel written during his own lifetime, Arnold Bennett's *A Man from the North* (1898). Richard Larch, a young clerk from the Potteries, moves to London to work in a solicitor's office and pursue his literary ambitions. One Saturday he has a chance meeting with Aked, a business colleague, at the British Museum Reading Room. Aked has gone to this classic Gissing location to borrow and re-read (not, we should note, to read for the first time) some of Gissing's novels. He is studying Gissing for a book he is planning, *The Psychology of the Suburbs*. Aked's book is never written, and nor is Larch's projected novel. The novel hinges on the relationship between a failed writer, his failed mentor, and the 'unsuitable' woman (a cashier in a restaurant) who loves both of them. The trope of the failed writer, whether fairly or not, long remained attached to Gissing himself. Pritchett, for example, would claim in an essay significantly entitled 'Grub Street' that the older novelist's fame was based on his 'failure and his exile' (Pritchett, p.62). Certainly it was a very different kind of fame from Arnold Bennett's.

A Man from the North was the obscure debut novel of a writer destined not only to become conspicuously successful, but to add to his own fame by celebrating that fame. A few years after *A Man from the North*, he himself would be writing guides for aspiring authors like Richard Larch. Bennett's literary trajectory not only suggests a more talented version of Gissing's own Jasper Milvain – 'Jasper of the facile pen' (*New Grub Street*, p.480) – but it defines a possible role for Gissing himself as symbolic literary precursor. Such a precursor would loom largest for novelists at the beginning of their careers, for novelists daunted by the struggle to realize their ambitions and writing to exorcise the fear of failure. He would exemplify the difficulty, the psychic cost and also the sheer drudgery of the professional writer's life. Aked lets Larch into the 'secret' of writing, a secret very Gissing-like in its suggestion of an inescapable circle of torment: ' "No writer does like writing, at least not one in a hundred, and the exception, ten to one, is a howling mediocrity. That's a fact. But all the same they're miserable if they don't write" ' (Bennett, 1912, p.81). As if this were not enough, the aspiring male writer has to cope with a tormented sexuality as well. Richard Larch has no success with middle-class women and, significantly, the most eligible person he meets – the daughter of a successful novelist who is herself an outspoken New Woman – holds no sexual interest for him at all.

Bennett's novel illustrates the fact that Gissing's life-story, and not merely his novels, was always present for the twentieth-century novelists who looked to him as some kind of precursor. This life-story had been the subject of literary gossip from the beginning, though curiously enough the first biography that appeared after his death was a thinly-veiled work of fiction, Morley Roberts's *The Private Life of Henry Maitland* (1912). In this roman à clef Roberts himself masquerades under the pseudonym 'J.C.H.', while Wells is 'G.H. Rivers', and so on. There is some amusement to be gained from the titles Roberts invented for some twenty of 'Henry Maitland's' books, beginning with *Children of the Dawn*, *Outside the Pale*, *Isabel*, and *The Mob*. Yet when he names the hero of *Paternoster Row* as 'Reardon' he gives the game away. There are long letters from Maitland to 'Rivers', some of which are authentic while others are possibly Roberts's inventions. Wells had written of Gissing's 'ill-advised ambition to write a series of novels' in his rejected preface to *Veranilda* in 1904 (Gettmann, p.265). Roberts picks up on the idea that Gissing should never have become a novelist. Maitland's 'true destiny', he tells us, 'was to have been a scholar at a great university' (Roberts, p.93). To make matters worse, his preface to *Henry Maitland* claimed that, though it was a 'sincere biography', certain facts about Maitland had had to be suppressed (p. vi). From a few heavy hints, we gather that Roberts believed, but did not want to say openly, that Gissing's whole life had been overshadowed by the syphilis that he caught from Nell Harrison. Maitland's death shows symptoms of tertiary syphilis (see Tindall, pp.152-155).

Roberts was not the first novelist to rework part, at least, of Gissing's life into fiction, and he would certainly not be the last. Wells, like Roberts, had rushed to their dying friend's bedside at St Jean Pied-de-Port in the last days of December 1903, and Wells drew on that experience in *Tono-Bungay* six years later to portray Edward Ponderevo, his patent-medicine king, raving on his deathbed. In our own time, George Gissing has appeared as a character in at least one historical novel of the late Victorian period, Peter Ackroyd's *Dan Leno and the Limehouse Golem* (1994). Gissing here is the promising young author of *Workers in the Dawn*. He spends his days in the British Museum Reading Room (sitting, so we are told, two seats away from the aged Karl Marx) and his nights trying to cope with his drunken wife. Nell's possible role as muse and evil genius seems inescapable in any future fictionalized account of Gissing's life; it has been fully explored by the novelist Gillian Tindall in her bio-critical study *The Born Exile*, despite being marginalized by Gissing's more academic biographers.

Gissing's personal life was barely summarized in the first critical monograph on the novelist, published in 1912 by Arnold Bennett's disciple Frank Swinnerton; but by general consent Swinnerton's effort stands even lower than Roberts' in the annals of Gissing scholarship. *George Gissing, A Critical Study* was commissioned for a series that also included volumes on Ibsen, Hardy, Swinburne, J.M. Synge and William Morris. Swinnerton seems to have intended a measured literary assessment of the recently dead writer, but what he actually produced was a laborious hatchet-job. His own early novels beginning with *The Merry Heart* (1909) are, as has often been noted, influenced by both Gissing and Bennett. His most successful novel was the melodramatic *Nocturne* (1917), where the setting varies from family bickerings in a cramped South London back-to-back to a lurid seduction scene in a yacht moored on the Thames at Westminster. The novel was published in wartime, and its lovers' parting after a one-night stand would have had a special meaning for wartime readers. Seasoned Gissing admirers, however, might have noted the 'sinister power' and 'almost uncontrollable terror' that the Thames waters under Westminster Bridge held for the heroine at the beginning (Swinnerton, 1917, p.19). Is she pregnant, and will she jump? Gissing, surely, would not have left us in any uncertainty as to the likely outcome.

Both Swinnerton and Roberts are novelists with a debt to Gissing, yet in their roles as Gissing's first critic and first biographer they did nothing to enhance his posthumous reputation. In the 1930s and 1940s it would have to be rescued by others, most notably George Orwell. Before moving on, however, it is appropriate to return briefly to Arnold Bennett. Like his character Aked, Bennett must certainly have read Gissing in the British Museum: Aked complains of the Reading Room's 'beastly' refusal to issue novels less than five years old (Bennett, 1912, p.72). Bennett's essay 'Mr George Gissing, an Inquiry', first published in 1899 and reprinted in *Fame and Fiction* (1901), deserves some of the classic status that Swinnerton and Roberts lamentably failed to achieve. Bennett approaches Gissing both as a consummate realist and as a recorder of class differences and social nuances: 'His philosophy seems to be that social distinctions have a profounder influence upon the general human destiny than is commonly thought' (Bennett, 1972, p.364). Gissing's attention to social distinctions appealed not only to Bennett but to Virginia Woolf, for whom the English novel would be 'unrecognizable' without the ups and downs of social rank (Woolf, 2003, p.214). Woolf believed that novelists, unlike painters or dramatists, were likely to be hampered by the limitations of their social experience, but part of Gissing's appeal to the twentieth century was that his biographical misfortunes forced him out of the grooves of middle-class fiction. As Bennett expressed it, 'The average artist stays at home in life; Mr Gissing has travelled far, and brought back strange, troublous tales full of disturbing beauty; and he suffers for his originality' (Bennett, 1972, p.365). This suggests that Gissing's true value as a precursor to modern novelists would lie in his alienated standpoint and in his record of the experiences of the social outlaw and the people whom he called the 'unclassed'.

The alienated intellectual

When he came to write a preface to the second edition of *The Unclassed* in 1895, Gissing had to acknowledge that his title had been misunderstood. To be unclassed was not to be déclassé – the term applied to an aristocrat who has come down in the world. Nor was it a euphemism for the 'condition technically represented by the heroine' – that is, prostitution. Instead, 'Male and female, all the prominent persons of the story dwell in a limbo external to society' (*The Unclassed*, p.v). This leaves it open whether or not to be unclassed is a voluntary state. The term distantly alludes to Matthew Arnold's men of culture, who stand apart from the three classes of aristocracy, bourgeoisie and proletariat, but in practice the 'unclassed' have either lost or repudiated the identity of the middle class whom Arnold denounced as Philistines. In *New Grub Street* Edwin Reardon's novels 'dealt with no particular class of society (unless one makes a distinct class of people who have brains)' (*New Grub Street*, p.54). Godwin Peak, the hero of *Born in Exile*, trains as an industrial chemist but aspires to the life of a well-to-do rentier. His 'unsocial' manners and 'repelling arrogance' are typical traits of a Gissing protagonist (*Born in Exile*, p.46). Such a protagonist is 'unclassed' in the sense of having failed to find what he considers is his rightful place in society. His gifts are unrecognized and he is condemned to poverty.

In *The Private Life*, Morley Roberts attributed to Henry Maitland the statement that 'the most characteristic, the most important, part of my work is that which deals with a class of young men distinctive of our time – well-educated, fairly bred, *but without money*' (Roberts, p.301). This would be echoed time and again by the novelists who wrote about Gissing. For Woolf, 'what Gissing proves is the terrible importance of money' (Woolf, 1972, p.531). Orwell summed up Gissing as 'the chronicler of vulgarity, squalor and failure' (Orwell, 1960, p.43), and called his first essay on his predecessor 'Not Enough Money'. At this point, we need to invoke Raymond Williams' term 'negative identification' in order to understand Gissing's characteristic attitudes towards money and class. What Williams had in mind was not the state of being unclassed as such, but the possibility, voiced only to be repudiated by Osmond Waymark in *The Unclassed*, that an intellectual alienated from the middle class will selfishly identify with the cause of the proletariat. According to Williams, Waymark's critical reaction against his earlier socialist views leads to an intense disillusionment and hatred of the proletariat which is equally based on negative identification. Ultimately it is a form of self-hatred, just as his love for the suffering masses was a disguised form of self-love.

The critique of negative identification alerts us to the self-destructiveness of Gissing's protagonists, and to their paranoid attitude to the people around them. It is this, rather than their unclassed status, that is Gissing's most significant fictional innovation. Other Victorian novelists such as George Eliot in *Felix Holt the Radical* (1866) and Walter Besant in *All Sorts and Conditions of Men* (1882) had portrayed characters who sacrifice personal advantages in order to pursue a philanthropic mission among the working classes. These characters overcome obstacles to enjoy an almost fairy-tale success, much to the satisfaction of the liberal middle-class reader. But any Gissing protagonist (like Egremont in *Thyrza*) who sets out with philanthropic ideas is headed for disillusionment and failure. Often the male protagonist starts out with no illusions and thus has no possibility of idealizing the poverty to which he finds himself condemned. All that he can do is to observe the life of the poor around him with assumed objectivity, since he believes that the people's condition is irreversible and that social reform would not improve it. This implies that there are two kinds of modern novel in which Gissing's influence may be detected, and is sometimes strongly marked. The first is the novel of the alienated intellectual who is, as Fredric Jameson suggests, Gissing's characteristic protagonist (Jameson, pp.195-201). The second is the novel of 'ordinary vulgar life' (*New Grub Street*, p.149) seen from an implicitly or explicitly alienated standpoint.

Gissing's joint concern with the alienated intellectual and with life below the level of the comfortable middle classes was by no means unique, though he was somewhat in advance of his English contemporaries. There would soon be other models for twentieth-century novelists to follow, James Joyce being by far the most prestigious. Joyce himself was no admirer of Gissing. He wrote to his brother in 1906 that he was 'labouring' to finish *Demos*: 'Why are English novels so terribly boring?... Without boasting I think I have little or nothing to learn from English novelists' (Joyce, p.518). There is no reason to disbelieve Joyce's assertion. Nevertheless, it is worth noting that he was a university graduate whose social experience was far more fortunate than Gissing's, even though both men spent their formative years as artists living in poverty. Stephen Dedalus could be the direct projection of a writer who had little to hide and who, in any case, eventually found private patronage to enable him to survive. Gissing, by contrast, both wrote for the market-place and had suffered a secret disgrace that could never be revealed to his respectable readers. He had been imprisoned, not for some political crime that might have been represented as heroic, but for a series of sordid thefts from his fellow-students in Manchester. This means that Gissing's best novels are the product of an artist with truths to express that could not be directly attached to his hero. *The Unclassed*, an early novel – Gissing described it as 'the work of a very young man, who deals in a romantic spirit with the gloomier facts of life' (*The Unclassed*, p.v) – provides a straightforward example.

Osmond Waymark lives a blameless life as an aspiring novelist, although his day job is that of a rent collector in the East End. We are not invited to look upon Waymark the rent collector as a kind of thief; instead, the motif of theft is dispersed throughout the novel and Waymark himself is kidnapped and robbed by a slum tenant. His working-class girl-friend Ida Starr is found guilty and imprisoned on a false charge of theft. His middle-class girl-friend Maud Enderby is the daughter of a clergyman unfrocked as a result of stealing from the poor-box. Clearly such a novel demands a degree of autobiographical reading, but it is very different from what we bring to a writer like Joyce.

In her 1927 essay on Gissing, Virginia Woolf addressed this question of distorted and unavowed autobiography in his fiction. His novels made for 'painful reading', since 'the people who suffer most are capable of making their suffering part of a reasoned view of life.... Their unhappiness represents something more than a personal reverse.' For Woolf, Gissing's intellectual quality allowed him to overflow the 'boundaries' of a limited personal existence, yet it fell short of the universal sympathy required of the 'great novelist'. Gissing's fiction remained 'solitary, self-centred, apart' (Woolf, 2003, pp.222f.). Something similar could be said of the characters in Woolf's own novels, such as Ralph Denham in *Night and Day* and Charles Tansley in *To the Lighthouse*, who strike us as Gissing-like. They are awkward, self-conscious outsiders and hangers-on of the intellectual aristocracy that Woolf tends to place at the centre of her fiction. Ralph Denham looking through the lighted windows of the Cheyne Walk house of Katherine Hilbery, the poised young hostess who is the granddaughter of a famous poet, is reminiscent of Godwin Peak (Woolf, 1969, p.366).

Gissing's protagonists are, as we have seen, not merely socially gauche; they also nurse a secret guilt. The twentieth-century novelists temperamentally closest to him were not just those who dreamed of failure, but those whose dreams (or acts) were tinged with guilt and self-destruction. They represent the novelist as alienated intellectual, whether or not their guilt and self-destruction were as palpable as Gissing's own.

From Jean Rhys to Philip Larkin

The theme of secret guilt leads inevitably to George Orwell: the Orwell who served as an imperial policeman and then threw up his career in disgust, the Orwell who posed as a down and out, the Orwell who hid his real identity under a pseudonym and whose greatest novel, *Nineteen Eighty-Four*, is riddled with pain and guilt. A full analysis of Orwell is clearly beyond the scope of this chapter, and all I can do is to hint at some of the insights that follow from his choice of Gissing as one of his literary mentors. In any case, Orwell at his most Gissing-like has curious similarities with a slightly earlier novelist of the 1930s, whose work he may not have known just as she herself may never have read Gissing. Jean Rhys's alienated protagonists, like Orwell's, reveal something of the Gissing hero's negative identification and 'double consciousness' (*Born in Exile*, p.71).

Rhys's novels introduce two plot-features that would become commonplace in twentieth-century fiction, though in Gissing they are of little importance. The first is the prominent use of childhood memories, while the second is an Oedipal drama played out in adulthood. In Gissing, memories of childhood, when they appear, are not so much a narrative obsession as an occasional device to fill in a character's background and keep the story going. As for family conflicts, Godwin Peak in *Born in Exile* abandons his university studies as a result of his uncle's seemingly random decision to open a café outside the college gates, but this is not the start of some epic struggle between uncle and nephew. Once Andrew Peak and his café have served their transient purpose, they are soon forgotten. Jean Rhys's early protagonists such as Julia Martin in *After Leaving Mr Mackenzie* (1930) and Anna Morgan in *Voyage in the Dark* (1934) are virtually paralysed by childhood memories, relived with an intensity that is in sharp contrast to the drabness and loss of purpose in their adult lives.

' "What I like about Gissing," ' says a feminist academic in *The Odd Woman* (1974) by the American novelist Gail Godwin, ' "is that he lets his characters think. They come to horrible ends, most of them, but they keep track of themselves so beautifully along the way" ' (Godwin, p.61). Rhys's early heroines are 'odd women' in this sense. Their guilty secret is that, though they remain obsessed with their own mothers, they are themselves the mothers of dead children. Anna Morgan has a backstreet abortion. Julia Martin's recollection is that her child, born to a failed marriage, died 'for the simple reason that [she hadn't] enough money to keep it alive' (Rhys, 1984, p.311). Julia, who has been living in Paris, comes back to England in time to witness her mother's death, and then quarrels passionately with her sister and her uncle. To them she is the black sheep of the family, though outwardly they all belong to the same shabby-genteel social stratum. It is her sister, not Julia herself, who, we are told, 'was labelled for all to see. She was labelled "Middle class, no money" ' (Rhys, 1984, p.289). The guilt of Rhys's protagonists is the result of their failed rebellion against the lower middle-class world to which they have, as it were, been born in exile. Not content to be 'members of the vast crowd that bears on its back the label, "No money" from the cradle to the grave' (p.279), they have drifted from one wealthy lover to the next in a process of increasingly explicit self-prostitution – but they still have no money. Anna in *Voyage in the Dark* lives in mean London streets that suggest both Gissing and the early Orwell. There is an aspidistra in the front hall of her lodging-house. Her room was previously occupied by a failed poet, who was evicted because of arrears of rent. Anna, however, not only alienates her friends by her unsocial arrogance but spends much of her time sleeping and daydreaming. Brought up in the Caribbean, she is literally an alien in London, unemployed, largely unknown and forever feeling the cold. It is her colonial background that has made her a spiritual aristocrat, regarding London's vulgarity with Peak-like disdain and expecting others to work for and support her. The less money she has, the more she despises 'The ones without any money, the ones with beastly lives.... They swarm like woodlice when you push a stick into a woodlice-nest at home. And their faces are the colour of woodlice' (Rhys, 1967, p.26). This might sound like a gold-digger's charter, but Anna's desire for money is ambivalent and shot through with negative identification. She thinks money should be for everyone, 'like water' (p.27). She despises it when she has it, and hates what she has to do to get it, but – as Orwell would write of the transaction between prostitute and client

– she goes ahead and does it just the same (Orwell, 1954, p.58).

The 'money-god' is Orwell's phrase, though Gissing or Rhys could easily have invented it. In Orwell's most Gissing-like novel, *Keep the Aspidistra Flying* (1936), Gordon Comstock's doomed and ineffectual rebellion against the money-god and its symbol, the aspidistra, is a manifest case of negative identification. Gordon has given up his job as an advertising copywriter to work as an underpaid bookshop assistant, and he is writing a poem, 'London Pleasures', 'a huge, ambitious project – the kind of thing that should only be undertaken by people with endless leisure' (Orwell, 1962, p.36). But if this is, like *A Man from the North*, the story of a failed writer – and one with a very similar outcome – it is also full of irrational, self-inflicted humiliations reminding us of the paranoia of a Gissing character such as Godwin Peak. But now the humiliations are no longer grandiose and potentially tragic, but petty and absurd. Gordon 'clung with a sort of painful joy to the notion that because he was poor everyone must want to insult him. It fitted in with his philosophy of life' (p.99). The incident of the 'Joey' or silver threepenny-bit at the beginning of the novel typifies the sheer triviality of Gordon's paranoia. Gordon's objections to spending his 'Joey' are mystifying to today's readers (and even to Orwell scholars). Gordon always knows he is being silly and childish, since the social humiliation he fears is either trivial or non-existent. Had Godwin Peak elected to stay at Whitelaw College once his uncle's café had opened, he would have faced some real embarrassments, and perhaps worse. Gordon, however, fantasizes that somebody would only try to spend a Joey if it was their 'last threepence in the world' (p.8): yet the Joey is legal tender and there is no obvious reason not to spend it. Since it is indeed his last threepence in the world, Gordon throws it away. It is the perversity of Gordon's rebellion against the money-god, and not just its miserable failure, that gives depth to *Keep the Aspidistra Flying* and accounts for its peculiar tone wavering between satire and painful confession. Orwell wrote that Gissing 'believed, perhaps not wrongly, that poverty causes more suffering in the middle class than in the working class' (Orwell, 1998, p.46). This certainly applies to Orwell's early fiction.

As I have already suggested, V.S. Pritchett's claim that Gissing may have been the 'perverse prophet and progenitor' of the 'grudge' novelists of the 1950s (Pritchett, p.63) should be read in the light of these novelists' known admiration for Orwell. Pritchett's comment is misleading to some extent, since novels like John Wain's *Hurry on Down* (1953) and Kingsley Amis's *Lucky Jim* (1954) are, among other things, witty romances owing as much, if not more, to the early Evelyn Waugh than to Gissing. Moreover, both Wain and Amis were university English lecturers given to self-conscious literary references, so that *Hurry on Down*, for example, ends inconclusively with the hero and heroine discussing the inconclusive ending of Moll Flanders. Where there is, it seems to me, a much more direct link to Gissing is in Philip Larkin's early fiction, which is known to have influenced both Amis and Wain. Introducing the 1964 reprint of his first novel *Jill* (1946), Larkin quoted James Gindin's suggestion that it 'contained the first example of that characteristic landmark of the British post-war novel, the displaced working-class hero'.

The displaced working-class hero is normally a 'scholarship boy' and university graduate, though Larkin's protagonist John Kemp is an undergraduate in his first term at Oxford. As such, he reminds us that before the Second World War students from the lower middle class, as well as the working class, would have needed a scholarship to go to university and would have been displaced from their social background by doing so. Possibly the earliest example of the fiction of the displaced undergraduate is the Whitelaw College section of *Born in Exile* (discussed in Jenny Bourne Taylor's chapter here). *Jill*, however, is an anarchic comedy in which the tone is set by Kemp's farcical behaviour on the train to Oxford to take up his university place. Travelling from the northern industrial city of Huddlesford, he goes to the toilet and throws his sandwiches out of the window because he cannot face the humiliation of eating them in front of his fellow-passengers. When he returns to the compartment, they all get their own sandwiches out while he is forced to go hungry. The incident is reminiscent of Orwell's Joey. Kemp's inability to read his social surroundings becomes much worse at Oxford, where he has to share his rooms with a wealthy middle-class lout. His lifelong paranoia now has a real justification: 'He was used to humiliating memories, but these seemed different, they had really existed' (Larkin, 1964, p.203).

Kemp finds some compensation for the humiliation and bullying he encounters at Oxford with a surge of imaginative invention, which turns him into a kind of novelist. To make himself more interesting he furnishes himself with a younger sister who has been sent to a private boarding school, Willow Gables. (Larkin himself wrote a pornographic romance about Willow Gables.) The story of Jill becomes bound up with Kemp's sexual obsessions, leading him to stalk a fifteen-year-old girl around Oxford. As his promising undergraduate career turns to dust, he directs his resentment against the self-righteous Whitbread, another working-class scholarship boy. In the novel's climactic scene Kemp vandalizes Whitbread's rooms, destroys his books and – perhaps to show that it is not a mere upper-class prank – steals a pound note from his jacket. This is surely a deliberate or unconscious echo of Gissing's biography.

John Kemp's imaginative alienation is that of an adolescent moral coward, not a mature intellectual: the novel ends when his father, a retired policeman, and his mother unexpectedly descend on the college. Larkin's second novel, *A Girl in Winter* (1947), portrays a very different kind of alienation since the protagonist is a German refugee (almost certainly Jewish, though this is not stated) working as a librarian in wartime England. A genuinely displaced person, she is shown painfully analysing and outgrowing her own negative identification (which began on a girlhood visit to the country before the War) with the English middle class. The middle classes in both novels are portrayed with a distaste verging on hatred. *A Girl in Winter*, even more than *Jill*, is the early work of someone who might have become a very fine novelist. But Larkin never took Gissing's crucial step of becoming a professional writer. The limitation of social experience that followed from his decision to remain as a full-time university librarian was clearly no handicap to him as a poet. For example, he could write a poem – not, of course, a novel – about not knowing Mr Bleaney (Larkin, 1971, p.10). How far Larkin's poetry expresses a Gissing-like mixture of intellectual alienation, negative identification and double consciousness is an interesting question, but it lies outside the scope of this chapter.

A fiction of shopkeepers

I have argued that Gissing's influence might be found in two kinds of modern fiction, the fiction of the alienated intellectual and that of ordinary 'vulgar' (that is, uneducated) life seen from an alienated standpoint. The latter is the domain of some of Larkin's most famous poems, but it remains comparatively rare in the novel, since fiction normally requires a strong element of romance to make ordinary experience attractive to readers. To follow in Gissing's footsteps would be to write of lower-class, non-literary lives without the wit and romance which, as Woolf noted, Gissing generally lacked. Add wit and romance, and the result might be (for example) Wells's comedies of draper's assistants and small shopkeepers, *Kipps* (1905) and *The History of Mr Polly* (1910). Take them away, and the best precedent, curiously enough, is a realist experiment carried out not by Gissing himself but by one of his characters, Harold Biffen in *New Grub Street*.

Biffen's book represents an alternative kind of serious novel to that produced by Edwin Reardon, the chronicler of alienated intellectuals or 'people who have brains'. Biffen's choice of the 'sphere of the ignobly decent' and his attempt to treat 'ordinary vulgar life with fidelity and seriousness' (*New Grub Street*, p.149) have, of course, various precedents and parallels in late nineteenth-century fiction; George Moore's *Esther Waters* (1894) is a well-known example in English. Biffen, however, projects an experimental or 'absolute' realism which, as he boasts, must be 'unutterably tedious' if it is to succeed: 'If it were anything but tedious it would be untrue' (*New Grub Street*, pp.149f.). The novel he decides to write is called *Mr Bailey, Grocer*:

> "Mr Bailey is a grocer in a little street by here. I have dealt with him for a long time, and as he's a talkative fellow I've come to know a good deal about him and his history.... Now I'm going to tell the true story of Mr Bailey's marriage and of his progress as a grocer. It'll be a great book – a great book!" (p.220)

We should note here how the educated novelist (Biffen is a classics tutor) depends for his material on an uneducated storyteller, a 'talkative fellow' who is represented as a habitual raconteur. If Mr Bailey read novels then Biffen would have no chance of publishing his life-story without acknowledgement. Biffen is the alienated intellectual as ghostwriter for the self-satisfied, unalienated citizen. *Mr Bailey, Grocer* falls dead from the press, as perhaps it deserves, yet it could be taken as a precedent for certain very famous twentieth-century novels which purportedly originated in chance encounters between the novelist and a potential subject. Virginia Woolf writes in her essay 'Mr Bennett and Mrs Brown' of the literary possibilities of an unknown woman, whom she names Mrs Brown, glimpsed in a railway carriage. Arnold Bennett claimed to have conceived *The Old Wives' Tale* (1909) as a result of a few minutes' observation of a clumsy old woman in a Paris restaurant (Bennett, 1911, p.v). James Joyce is alleged to have spent part of a single day in the company of Alfred Hunter, the original of Leopold Bloom. The interior monologue as practised by Joyce, Woolf and their contemporaries proved to be the most appropriate method for constructing a literary fiction on the basis of unsophisticated, oral storytelling by the principal characters.

Harold Biffen himself is a stereotypical alienated intellectual. He lives on his own in a rented room. He would rather starve than take a regular job, and his final gesture, after the failure of his novel, is to commit suicide. How was his negative identification with Mr Bailey (a man for whom he evinces, if anything, a mild disdain) reflected in his fiction? We do not know the answer to this, but twentieth-century novelists using the interior monologue have generally granted their characters a degree of linguistic articulateness much superior to what would be expected in actual life. The novelist as ghostwriter gives the character's inarticulate or semi-articulated thoughts shape and form; but this is all too easily confused with the novelist as ventriloquist putting strange words into the character's mouth. In Orwell's *Coming Up for Air* (1939), for example, the first-person narrator George Bowling tends to 'Orwellise' (as one critic has put it) even though he is represented as an insurance salesman devoid of literary or intellectual pretensions (Zwerdling, p.148).

At the core of *Coming Up for Air* is a romantic yet disillusioned return to the scene of Bowling's childhood. This suggests that the modern avatars of *Mr Bailey, Grocer* might be novels in which 'ordinary vulgar life', or the 'sphere of the ignobly decent', is represented by the world of the alienated author's parents. Childhood memory and Oedipal drama are thus incorporated into a dispassionate yet deeply obsessive account of a social milieu from which the author has become detached by his or her experiences of higher education. John Wain in *Hurry on Down* offers the following passing vignette of a working-class scholarship boy's rejection of his parents:

> The raw cult of success by which he lived could allow of no tolerance of the couple who had spawned him; they were neither prosperous nor celebrated, their Birmingham speech exposed in an instant the unreality of his own diction (an unbelievably exact reproduction of Lockwood's donnish snivelling), and, in short, his resentment of them could go no further. (Wain, p. 14)

Perhaps we should not be surprised that Hutchins, the character concerned, ends up by committing an act of theft (to pay for his girl-friend's abortion) which he successfully blames on Wain's middle-class protagonist – yet another possible recycling of Gissing's life-story. But there is a different kind of novel in which, prompted perhaps by the death of one or both parents, the priggish scholarship boy or girl discovers that they are destined to be a writer and that the parents' provincial and ignoble life is the writer's unavoidable subject. V.S. Naipaul has hinted at such feelings in various memoirs and in the character of Anand in *A House for Mr Biswas* (1961). Yet Mr Biswas, like Naipaul's own father, was a successful journalist and to some extent a failed novelist. Biffen's Mr Bailey, by contrast, is someone we should picture as being entirely reconciled to his life as a grocer.

In twentieth-century English fiction there is a small but continuous tradition of novels of shopkeepers and shop assistants: not enough to justify Napoleon's famous observation, but enough, perhaps, to suggest Harold Biffen as a possible precursor. Yet many of these novels are melodramatic romances, including Joseph

Conrad's *The Secret Agent* (1907) and Arnold Bennett's *Riceyman Steps* (1923), both of which may have been inspired by sensational newspaper reports. The quite unsensational Gordon Comstock was, however, a bookseller's assistant. Among very recent writers, Graham Swift launched his career with a work which, of all the shopkeeping novels, seems closest to the Mr Bailey ideal. *The Sweet Shop Owner* (1980) uses a form of interior monologue to record the last day in the life of Willy Chapman, who has run a newsagent's shop in South London ever since getting married some 35 years earlier. Like Gissing's Mr Bailey, he had no money of his own but married a strong-minded woman who gave him the money to start up his business. Swift's first novel has all the qualities of his later, more celebrated works such as *Last Orders* (1996), and his exploration of the 'psychology of the suburbs' is outstandingly successful. Necessarily he has to endow Willy Chapman with a novelist's eloquence, though Chapman never even bothers to read the newspapers that he sells, let alone put pen to paper. Chapman is a widower who eventually dies of a heart attack and must have been found alone in his house, possibly, we conjecture, by his daughter. And it is Dorothy, the educated daughter, who seems to be the source of the novel's alienated standpoint. Among Chapman's social circle she is the only person who could possibly share the novel's intense, archaeological interest in reconstructing every detail of the shopkeeper's last day alive, though given what we are told of Dorothy this does not seem very likely.

Swift's second novel *Shuttlecock* (1981) is an Oedipal drama involving melodramatic discoveries and the metafictional deconstruction of a romance narrative written by the protagonist's father. Starting out in a George Bowling-like vein, *Shuttlecock* turns into an investigative thriller. There is no such romance element in *The Sweet Shop Owner*, which remains austerely true to its premise of portraying what Gissing would have called an ordinary vulgar life. But Swift's first novel begins with Chapman receiving a coldly contemptuous letter from Dorothy which is, apparently, meant to sever all relations between them. Dorothy, born in 1949 (the same year as Swift, as it happens), has transcended her father's world by taking a First in English at London University, where she began writing a thesis on 'Romantic Poetry and the Sense of History'. Now living with a university lecturer in Bristol, she acknowledges no remaining ties with her own family. Will Dorothy return to discover her father's dead body? We do not know, but she will presumably have to attend his funeral, which may be enough to convey her father's last unspoken message to her: that, try as one may, one cannot escape or outlive one's own history. Everything in Willy Chapman's last day, including his death, expresses a single longing, which is to explain and justify his life in Dorothy's eyes. Swift's achievement in *The Sweet Shop Owner* is to find a form in which the fiction of ordinary suburban life, far from being unutterably tedious, becomes full of poignancy and suspense at the same time that – by implication at least – it acknowledges the alienated intellectual's standpoint. The intellectual herself can never appear, though her place is to some extent taken (as in Swift's other novels) by ideas of history that are stated in the narrative, and by the figure of the history teacher as disregarded mentor.

Graham Swift's fiction suggests one way in which Gissing's penumbral presence might be destined to continue in the twenty-first century. However much we might wish that social mobility, mass higher education and general prosperity had made Gissing's class-dominated world a thing of the past, his fiction cannot be reduced to costume drama or to a source of picturesque 'heritage' experience. So long as differing levels of class and educational opportunity remain inescapable factors in our society, there will be a possibility of being 'unclassed' and a desire to hear the voices of those so positioned. Does this mean that Gissing will still be seen (to some extent, at least) as a misguided and failed artist and a perpetual outsider? Virginia Woolf's final verdict on him in her 1927 essay – that he was an 'imperfect novelist, but a highly educated man' (Woolf, 2003, p.225) – is not quite as crushing as those of Wells and Roberts, but it is unequivocal in excluding him from what F.R. Leavis would soon be calling the Great Tradition of English fiction. Gissing is a novelist with a hidden grievance, Woolf seems to be saying, and his sympathies are not universal enough.

To circumvent this judgement we need to re-emphasize that Gissing represents the novelist as alienated intellectual, as Joyce also does. In recent years Joyce's alienation has been widely understood (and perhaps sentimentalized) as that of a colonial subject who was forced to write and speak in the acquired language of the colonizer. Stephen Dedalus, fired by this insight, dreams of becoming the conscience of modern Ireland, but Joyce's own alienation, which was rather different from Stephen's, led ultimately to *Finnegans Wake* and the cosmopolitan babble of a writer playing language-games in a near-vacuum. Joyce no more fits Virginia Woolf's idea of the perfect novelist than Gissing does – and, even if Joyce had done, he was not content to remain so: but Gissing, who always had a living to earn, remained wedded to the conventional fictional structures that Joyce rejected. If Joyce is the 'fabulous artificer', Gissing might be understood as a kind of furtive artificer, constructing his fictions as a criminal constructs alibis and false leads, playing games with the novel-buying public in order to keep writing novels while he himself remains hidden. Not even Edwin Reardon, Gissing's portrait of the intellectual novelist, resembles his creator as closely as Stephen Dedalus resembles the young Joyce.

At least one critic, Fredric Jameson, has detected a deeply unsettling quality in the very language of Gissing's fiction. Language in the early Gissing novels, Jameson suggests, undergoes reification and becomes a 'virtually Flaubertian mimicry of received ideas', while Gissing's later style converts literary English into 'a dead language like Latin' (Jameson, pp.202f.). Gissing's negative identification, in Jameson's view, leads not just to self-destructive and paranoid reactions to the social world but to a kind of linguistic maladjustment, an inability to feel at home in the language of his contemporaries which is, once again, symptomatic of his 'unclassed' position. If this is so, Gissing's successors have sought, by and large, to drag his work back to the mainstream, diluting what Woolf called his 'peculiar grimness' with the humour and sentiment inherent in her humanistic view of the perfect novelist (Woolf, 2003, p.223). This strategy on the part of twentieth-century novelists is readily understandable; without it, Gissing's work might have been altogether forgotten. But it means there are resources of radical alienation in his fiction which remain untapped, and perhaps even undetected. They are there for future novelists to investigate.

Bibliography

Ackroyd, Peter (1998), *Dan Leno and the Limehouse Golem*, Vintage, London.
Allen, Walter (attrib.) (1968; first publ.1948), 'The Permanent Stranger', unsigned article in Times Literary Supplement , repr. in Coustillas, P. (ed.), *Collected Articles on George Gissing*, Cass, London.
Bennett, Arnold (1911; first publ. 1909), 'Preface' to *The Old Wives' Tale*, Hodder, London.
Bennett, Arnold (1912), *A Man from the North*, Methuen, London.
Bennett, Arnold (1972; first publ. 1899), 'Mr George Gissing, an Inquiry' , in Coustillas, P. and Partridge, C. (eds.), *Gissing: The Critical Heritage*, Routledge, London and Boston, pp.361-365.
Gettmann, Royal A. (ed.) (1961), *George Gissing and H. G. Wells*, Hart-Davis, London.
Gissing, George (1958; first publ. 1891), *New Grub Street*, Oxford University Press, London.
Gissing, George (1985; first publ. 1892), *Born in Exile*, Hogarth, London.
Gissing, George (1895) (second edn.), *The Unclassed*, Lawrence and Bullen, London.
Godwin, Gail (1980), *The Odd Woman*, Warner, New York.
Jameson, Fredric (1981), *The Political Unconscious: Narrative as a Socially Symbolic Act*, Methuen, London.
Joyce, James (1972), 'James Joyce on Gissing', in Coustillas, P. and Partridge, C. (eds.), *Gissing: The Critical Heritage*, Routledge, London and Boston, p. 518.
Kirkpatrick, B. J. (1980: third. edn.), *A Bibliography of Virginia Woolf*, Clarendon, Oxford.
Larkin, Philip (1964: second edn.), *Jill: A Novel*, Faber, London.
Larkin, Philip (1971), *The Whitsun Weddings*, Faber, London.
Orwell, George (1954), *Nineteen Eighty-Four*, Penguin, Harmondsworth.
Orwell, George (1960), 'George Gissing', *London Magazine*, vii: 6, pp.36-43.
Orwell, George (1962), *Keep the Aspidistra Flying*, Penguin, Harmondsworth.
Orwell, George, (1998; first publ. 1943), 'Not Enough Money: A Sketch of George Gissing', in *The Complete Works of George Orwell*, Volume 15, ed. Peter Davison, Secker, London, pp.45-47.
Pritchett, V.S. (1965), 'Grub Street', in *The Working Novelist*, Chatto, London, pp.62-67.
Rhys, Jean (1967), *Voyage in the Dark*, Deutsch, London.
Rhys, Jean (1984), *After Leaving Mr Mackenzie*, in Jean Rhys, *The Early Novels*, Deutsch, London, pp.249-357.
Roberts, Morley (ed.) (1923: second edn.), *The Private Life of Henry Maitland*, Eveleigh Nash, London.
Swinnerton, Frank (1912), *George Gissing: A Critical Study*, Secker, London.
Swinnerton, Frank (1917), *Nocturne*, Hutchinson, London.
Tindall, Gillian (1974), *The Born Exile: George Gissing*, Temple Smith, London.
Wain, John (1960), *Hurry on Down*, Penguin, Harmondsworth.
Williams, Raymond (1961), *Culture and Society 1780-1950*, Penguin, Harmondsworth.
Woolf, Virginia (1969), *Night and Day*, Penguin, Harmondsworth.
Woolf, Virginia (1972; first publ. 1912), 'The Novels of Mr George Gissing', in Coustillas, P. and Partridge, C. (eds.), *Gissing: The Critical Heritage*, Routledge, London and Boston, pp.529-534.
Woolf, Virginia, ed. Andrew McNeillie (2003), *The Common Reader: Volume II*, Vintage, London.
Zwerdling, Alex (1974), *Orwell and the Left*, Yale University Press, New Haven and London.

Index

The Works of George Gissing

General Index

For Product Safety Concerns and Information please contact our EU
representative GPSR@taylorandfrancis.com
Taylor & Francis Verlag GmbH, Kaufingerstraße 24, 80331 München, Germany

www.ingramcontent.com/pod-product-compliance
Lightning Source LLC
LaVergne TN
LVHW021109300926
849160LV00009B/660

* 9 7 8 1 1 3 8 6 7 5 5 3 7 *